SOILS, LAND, AND LIFE

SOILS, LAND, AND LIFE

Stanley W. Buol

Professor Emeritus
North Carolina State University

PEARSON

Prentice
Hall

Upper Saddle River, New Jersey 07458

Library of Congress Cataloging-in-Publication Data

Buol, S. W.
 Soils, land, and life / Stan Buol. — 1st ed.
   p. cm.
 Includes bibliographical references and index.
 ISBN-13: 978-0-13-191481-0 (alk. paper)
 ISBN-10: 0-13-191481-2 (alk. paper)
 1. Soil management. 2. Land use. 3. Food. I. Title.
 S591.B8874 2008
 631.4—dc22

                              2007031488

Editor-in-Chief: Vernon R. Anthony
Associate Managing Editor: Christine Buckendahl
Editorial Assistant: Yvette Schlarman
Marketing Manager: Jimmy Stephens
Production Liaison: Joanne Riker
Cover Design Director: Jayne Conte
Cover Design: Bruce Kenselaar
Cover Photo: Getty Images, Inc.
Full-Service Project Management/Composition: Integra Software Services, Ltd.
Printer/Binder: Courier Companies, Inc.

Credits and acknowledgments borrowed from other sources and reproduced, with permission, in this textbook appear on appropriate page within text.

Pearson Education LTD.
Pearson Education Singapore, Pte. Ltd
Pearson Education, Canada, Ltd
Pearson Education — Japan

Pearson Education Australia PTY, Limited
Pearson Education North Asia Ltd
Pearson Educación de Mexico, S.A. de C.V.
Pearson Education Malaysia, Pte. Ltd

10 9 8 7 6 5 4 3 2 1
ISBN-13: 978-0-13-191481-0
ISBN-10:   0-13-191481-2

CONTENTS

PREFACE

As the population of the world has become more urbanized, people have become remote from the source of their food. Many urban dwellers see the land on which their food is grown only from the interstate highways or from the window of an airliner. Some remember early childhood experience on grandparents' farms and are disturbed by the large fields and mechanized methods of farming that now occupy the sites of those childhood memories. News items on television and in the print media report findings of contaminated water, suspected food poisoning by pesticides, destruction of tropical forests by shifting farmers, and the starvation of children around the world. Major news events where fertilizer is used to build bombs and localized events of contaminated food often cause the city dweller to suspect that all is not well on the land and to advocate a return to the more so-called natural systems of food production.

Why has farming in the Western world adopted the use of fertilizer and pesticides? Why are the farmers in tropical jungles cutting and burning the trees? Is it ignorance, greed, or necessity that has brought about such land use? Could it be that the soil is being destroyed? Or could it be that new technology has replaced older technology and in so doing solved old problems and replaced them with new problems?

This book addresses food production methods throughout the world by presenting the basic physical and chemical dynamics of the human food chain. Natural resources of soil, water, and temperature are not equal throughout the world. Land areas serve as solar panels where sunlight energizes the crop plants to convert the chemical elements necessary in human diets into organic compounds that can nourish the human body. Almost no food is consumed at the site where it is grown. Chemicals contained in food are thus transported by humans and concentrated as waste and potential pollution in cities throughout the world. Whether the city, as it were, is an extended family compound in a tropical jungle or an urban megalopolis, only the scale of chemical transport differs. The cycle of chemical elements necessary for life from soil and back to soil from where they originated is not often completed where human food production is involved. Several scenarios of food production, each compatible with ambient social, political, and economic realities, have developed to assure continued flow of food from the soil on the land to the mouths of humans.

The subjects of soil, land and life are presented in a form that should be easily understood by high school graduates. To those with science backgrounds the treatments will be almost too simple. No mathematics beyond adding, subtracting, and occasionally multiplying and dividing is used. All the chemistry is treated from a mass balance of elements in food production and consumption. The chemical transformations are described in a qualitative context. Similarly, the physics of water in the soil and landscapes is treated in an arithmetical fashion.

ACKNOWLEDGMENTS

Special thanks is extended to E. J. Kamprath, D. L. Hesterberg, and D. K. Cassel for reviewing several portions of the manuscript; Robert Austin and Gregory Buol for technical help in the preparation of several figures, and especially to Roberta Miller-Haraway for drafting several of the figures and preparation of all the figures for publication. Except where a source is credited, all figures originated with me, redone portrayals of well-known concepts the origins of which have been lost, or from duplicates of slides given to me by friends for classroom use over many years. To those friends I express gratitude and apologize for not giving credit.

To a multitude of unknown individuals throughout the world who have afforded me the opportunity to observe their daily activities, I owe an enormous amount of gratitude. These are the unnamed farmers, merchants, and laborers who plant, cultivate, harvest, and market the food supply of the world. From observing their actions I have attempted to assemble and collate the elements of nature from which people formulate the practices necessary to sustain human life. Many of my most appreciated teachers can neither read nor write, and few could communicate directly with me because of my limited linguistic ability. Regardless of language, all successful farmers in the world daily utilize the functional interactions of physics, chemistry, meteorology, botany, and zoology within a multitude of social, political, and economic conditions as they practice the art of farming. Their practical abilities defy academic capture.

ABOUT THE AUTHOR

Stanley W. Buol is a Distinguished Professor Emeritus in the Soil Science Department at North Carolina State University in Raleigh. He was reared on a farm near Belleville, Wisconsin, and completed his undergraduate and graduate work in soil science with a minor in meteorology at the University of Wisconsin-Madison. In 1960 he accepted a position as assistant professor in the Agricultural Chemistry and Soils Department at the University of Arizona in Tucson. In 1966 he joined the Soil Science Department at North Carolina State University where in 1969 he was promoted to professor and in 1992 received a William Neal Reynolds Distinguished Professorship. His primary responsibilities were research into the formation and classification of soils and graduate instruction. He served as major professor for 42 Ph.D. and 35 M. S. students and in 1990 was honored with an Alumni Distinguished Graduate Professorship. Many of his graduate students conducted research in North Carolina and Arizona, and several conducted extended field studies in South and Central America, Africa, and Asia. While supervising graduate students and cooperating with several national, international, and nongovernmental organizations, Buol traveled to and studied soils in over 60 countries, primarily in the tropical areas of the world. He received the International Soil Science Award from the Soil Science Society of America in 1989 and their Distinguished Service Award in 2005.

In addition to numerous research publications, experiment station bulletins, and book chapters, Buol has coauthored five editions of the graduate text *Soil Genesis and Classification,* which has been translated into four foreign languages.

CHAPTER

1

Introduction

We see plants and animals and recognize each as life-forms. We see hills and valleys and recognize land as a legal entity that can be bought and sold. Intuitively, when we see trees, grass, or flowers on the land, we know their roots are in something called soil. We seldom see more than the surface of that soil. Often we see pieces of soil, best referred to as dirt that sticks to our shoes and dulls the polish we work so hard to put on our cars. Dirt gets under our fingernails.

The association of life and land with the underlying soil is so common to human experience that we accept news broadcasts stating, "The forest fire destroyed 10,000 acres (4,049 hectares [ha]) of land" that in reality means "The forest fire destroyed the vegetation over 10,000 (4,049 ha) acres of land." The land remains after the fire. The soil under that land surface also remains after the fire. Although it may be somewhat changed by the fire, the soil is capable of supporting a new generation of plant life on the land. Although the spatial association of land, life, and soil is easily recognized, the temporal relationships are less well recognized. We are well acquainted with the life cycle of most plants and animals. We see seedlings emerge from the ground. We see leaves form on growing plants, the formation of flowers, followed by seeds, and accept that death will take place. Land and soil are much more permanent to the human experience, but they also change with time. The life cycle of land and soil usually exceeds that of life-forms, and changes are either so slow or sporadically rare that within our life span we tend to believe land and soil are permanent, or at least best considered in the context of a geologic time scale.

Soil is a complex media with a wide range of dynamics. Soil is both an inorganic entity composed of minerals that tend to be altered in a geologic time scale and an organic entity composed of dead and living life-forms that respond to biological time scales. Soil also has meteorological dynamics of temperature, which change from sunrise to sunset and from season to season, and moisture dynamics, which change in response to weather conditions. Thus it can be said that soil is both "as old as the hills" and "as changeable as the weather." Soil not only contains visible life-forms such as plant roots, insects, and worms but also millions of unseen microbes per cubic inch.

Human societies historically have sought land that provided for human necessities of food and fiber. They developed numerous and diverse cultural practices to wrench these necessities from the land they inhabited. Sites suitable for early human habitation were primarily dictated by climates with temperature and moisture suitable for producing human food. The geographic distribution of such climates shaped the development of early civilizations. Although temperature and moisture were readily apparent to human intellect, the complexities of the chemical requirements of life remained obscure for many human generations. Only during approximately the most recent 150 years have scientific investigations revealed an understanding of the chemical interactions of soil and life. Whereas past generations sought the chemical necessities of life by trial-and-error exploration of various lands in search for reliable food production, it is now possible to sustain and enhance the production of human food by augmenting the natural concentration of chemicals in the soil. The application of technologies to enhance sustained human food production is presently very uneven throughout the world. The technologies of enhanced and sustained food production may appear relatively simple, but their successful application involves the integration of technical, economic, and political entities within human societies. Modern societies pursue the quest for sustainable human survival by utilizing a multitude of political, economic, technical, and cultural strategies to gain human sustenance from the land and soil. Although many societies have been very successful in enhancing human food production per unit of land, many have failed. The failure of societies to produce food without adequately protecting the soil has resulted in devastating erosion in many parts of the world throughout recorded history (Lawdermilk, 1953). These sad experiences are avoidable only if all segments of society from political leaders to peasant farmers function in harmony to create and sustain viable agricultural systems that assure adequate nourishment to all within that society.

BOX 1-1

Evidence at the Scene of the Crime

Erosion is a natural process on all land areas of the earth. However, erosion caused by humans that degrades the ability of soil to provide an adequate foundation for future generations to grow food can be considered a crime. At every crime scene there is some evidence, but that evidence may not clearly identify the perpetrator of the crime. Evidence at a crime scene may be masked, removed, or overlooked by investigators. Investigations of why early civilizations failed have often concluded that excessive erosion was the criminal. However, a multitude of evidence would support the following statement: as written by the Soil Survey Staff of the U.S. Department of Agriculture.[1] "In fact, if continuing cultivation is assumed, generally, although not always, low fertility can be regarded as a main cause of erosion."

Using this statement as a premise to more fully investigate the crime of soil erosion, we must consider that ancient civilizations had no knowledge of chemistry. Upon their first settlement of land where good crops could be grown, the civilization flourished and populations increased. Cities were built, and food grown in the surrounding land was transported to the cities. That food contained the chemical elements necessary to life. Those elements were gradually removed from the soil where the food was grown. As the soil

(continued)

(Continued)

became impoverished, the crops grew less vigorously and failed to protect the soil from the erosive energy of rain and wind. Erosion resulted, and after the soil was removed no evidence of the diminished chemical composition of that soil remained. Only the evidence of erosion in the form of gullies, shallow soils, and bare rocks remained as clear evidence of erosion. Investigators then concluded that erosion was the cause. But in reality had erosion only been an accomplice that removed the evidence of the real culprit, soil with diminished fertility? As you read this volume, you can be a member of the jury and form your own opinion.

[1]Soil Survey Staff. 1951. *Soil Survey Manual.* USDA Handbook No. 18. Washington, DC: U.S. Government Printing Office, p. 261.

WE EACH KNOW SOME LAND, LIFE, AND SOIL SYSTEMS

Each of us has a personal understanding of an area of earth's surface. Each of us traverses some land each day. The land we see may be covered by vegetation, buildings, roads, or parking lots supported by soil. Our individual understanding of environments or ecosystems has been formed from these commonplace experiences. We often hear about soil in the context of soil erosion or soil pollution integral to environmental concerns. Sediment in our lakes and streams comes from soil erosion. Contaminated soil is removed from toxic waste spills. Landslides are soils rushing down hillsides, often demolishing buildings and roads (Figure 1–1). Dust in the air is

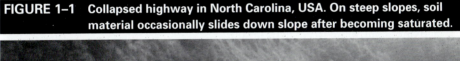

FIGURE 1–1 Collapsed highway in North Carolina, USA. On steep slopes, soil material occasionally slides down slope after becoming saturated.

soil particles lifted by the wind. There is an awareness that our agricultural production comes from the land and is somehow related to the fertility of the soil. We may attribute land covered with abundant fields of grain as having fertile soil. When our lawns or flowers do not fulfill our expectations, it is common to associate their ills with a soil condition. Perhaps we need to irrigate or fertilize the soil. If we investigate by digging in the area of unsatisfactory plant growth, we always seem to find the soil too sandy, too rocky, too hard, or too clayey. If we are successful in improving our lawns or flowers with some soil amendment or cultivation practice, we supposedly have a "green thumb" and become so-called soil experts, for that soil, for that type of plant, for that year.

Each of our individual experiences with land, life, and soil systems is most often limited to relatively small areas of land. Some people move to other areas of the world and find that cherished recipes for dealing with soil may or may not produce the same results at the new location. A move of only a few miles may reveal several contrasts in soil conditions. Farmers recognize differences in the soil properties within their farm and compensate for these differences by using different land management practices on each contrasting parcel of land. Football and baseball players recognize the different feel of the soil on different playing fields. Soil and land are mosaics most often covered by vegetative life. Some land features, such as hills and plains, and soil features, such as color and texture, are easily recognized by most people. Many soil features critical to its ability to grow plants are detectable only by scientific measurements of chemical and physical properties. More significantly, many properties of land and soil are not constant but change in response to weather and human activity. It is often said that no two years are exactly alike with respect to weather. In a similar vein, no two soils are exactly alike in chemical and physical composition. Further, the same soil changes in many respects from day to day and from year to year. Both natural events and human activities contribute to changes in soil.

To conceive of soil, land, or life as one thing is as noninformative as to conceive of all cars as "the car." Many features of all cars are similar and, in fact, the same from one car to the next, just as there are many properties of soil, land, and life that are the same or very similar in all kinds of soil, land, and life. However, just as cars differ in color, weight, horsepower, and tire size, soil, land, and life-forms differ in a nearly infinite number of details. Many driving skills and maintenance practices apply to all cars, but to obtain maximum performance, economy, and longevity, these practices differ from one model of car to another. Driving practices differ from crowded city streets to open interstate highways. In like fashion, human practices to acquire food and fiber differ in response to both properties of the land and soil and prevailing social and economic conditions.

Various chapters in this book discuss the human management requirements for different soil, land, and life systems. It is important to recognize that management requirements for food production differ depending on the properties of the land and soil. We must also recognize that cultural practices used to acquire human food and shelter depend not only on the soil and land conditions but also on the contemporary structure and expectations of the human society surrounding the tillers of the land, that is, the farmers. For example, where there are no alternative sources of food and little or no reliable market for food products, the goal of a farmer may be to provide a reliable food supply only for the immediate family (i.e., subsistent farmers). In other places and times when societies provide adequate infrastructure, farmers may specialize in the efficient production of only one commodity that can be reliably sold to secure enough money to purchase food and shelter for the family.

SETTING OF THE SOIL, LAND, AND LIFE DRAMA

The drama of soil, land, and life takes place where sunlight strikes the surface of the land. Powered by the energy of the sun, plants gather inorganic forms of life: essential elements from gases in the air above the land and minerals in the soil and water to form organic compounds that are suitable foods for animal life. Upon the death of plants or animals that consume the plants, these organic compounds decompose and the life-essential elements they contain return via various pathways to inorganic compounds in the air, water, or soil. Life-essential elements can neither be created nor destroyed, and only a finite quantity of each life-essential element is present on earth. In their quest for food, humans drastically alter the natural return of life-essential elements to the soil where it was produced as they transport life-essential elements in the form of food from the land where human food is grown to sites of food consumption.

The soil in which this drama is enacted is a dynamic and revolving stage. It heats as the sun rises in the morning and cools as radiation escapes to the air and space beyond at night. Soil becomes soggy when saturated with water after a rain and contains almost no water after rainless months during which plants have extracted water from the soil. In some parts of the world, the coldness of the winter causes the water in the soil to freeze and living organisms to become dormant. In other parts of the world, the cold of winter never comes, and in other lands, parts of the soils are permanently frozen. These are but some of the dynamic physical and biological processes that determine the interactions of soil, land, and life.

As we can observe, no two performances of any stage play are exactly alike. So it is with the performance of life and soil. A photograph can be taken to document a scene in a stage play, a flowering plant can be photographed, and a sample of soil can be taken to document the soil condition at one point in time, but a still photo or soil sample is not capable of capturing the dynamics of the entire performance. The appearance and activities of vegetation and other organisms differ from place to place and season to season. Climate contrasts from the poles to the equator and from sea level to mountaintop, and weather changes from day to day. The mineral components and thus the content of many essential elements in the soil differ from place to place.

Does each of us know life? Certainly we each know some forms of animal and plant life, but few of us have experience with microbial life. Does each person know land? Certainly we each see the hills and valleys we traverse, but few have experienced all lands. Does each person know soil? I think so. We each know soil in the context of our experience. Does each person know soils? Perhaps a few soils, but no one person has experience with all soils. It is inconceivable that any single person could achieve experience with the daily and annual dynamics present in each kind of soil in the world. Does someone have knowledge of each kind of soil and life in the world? Probably yes.

Whether through farming, mining, or building, human activity comes in contact with soil and almost all forms of life. Wherever humans reside and seek sustenance from plants growing on the soil they, by necessity, become well acquainted with a few kinds of the soil, land, and life-forms at that site.

Human attempts to communicate information relating to soil and land become difficult and misunderstandings are common because each person has personal concepts based on experience with only a small range of soil properties and life-forms present in and on the land areas of earth. Communications are further hampered by each individual's limited knowledge and experience with the various political, economic, and social environments that affect how humans respond as they meet the challenge of acquiring food from the locally available land and soil resources.

Each of us is likely to have a somewhat different concept of soil, land, and life. The following concepts of soil, land, and life are used in this volume.

Land is the solid portion of the earth surface not covered by ice or water. Land implies the presence of a quantity of air over the land. Land area is the basis for evaluations of life. For example, we speak of population density as number of people per unit of land area such as a square mile or square kilometer. Plant yields are expressed as tons of hay or bushels of corn per acre or hectare. Land also has form that we can see. Flat land connotes level areas of the earth surface. Such terms as *hilly land* or *mountainous land* are commonly used to express degrees of vertical variations in the earth surface. Land is also used to identify political boundaries and areas where societies flourished, such as the homeland. Land is a legal entity spatially defined in a deed of ownership. Land is a commodity that can be valued in monetary terms such as dollars per acre or dollars per house lot. Land is often identified according to human activities, such as game land for hunting or urban land where we build our cities. Concepts of land intuitively include the soil under the land surface and the air over the land.

Perhaps the most significant contribution of land as a factor in supporting life is the interception of sunlight. Radiant energy from the sun is required for plants to reassemble inorganic elements from the air, water, and soil into organic compounds that provide food for animals including humans. Without radiant energy from the sun, life as we know it could not exist. Only a fixed amount of radiant energy is available per unit area of land, and that amount differs depending on the geographic location of the land with respect to exposure to the sun as the spinning earth annually rotates around the sun. At any location the amount of land area determines the amount of radiant energy that can be captured.

Soil is a volume of material, bounded on top by the surface of the land and extending downward to include all material that is capable of supporting plant roots. Soil is not owned or politically claimed except as a component of the land within which the soil is fixed is owned or claimed. Soil must contain plant roots or be capable thereof. Recognizing that plants greatly expand their root systems as they mature from a seedling to a mature plant, we must consider "capable of supporting plant roots" as part of a definition of soil, although at any specific time no roots may be present. Also, different species of plants have different rooting depths. For these reasons it is not possible to assign an exact depth as a parameter of soil, although 80 inches (2 meters) is considered adequate to classify soil for most soil–plant interactions.

Soils have four vital functions in supporting life:

1. Soils physically hold plants upright, enabling them to access sunlight.
2. Soils have temperatures compatible with the growth of living organisms.
3. Soils store water in quantities required by plants between rainfall events.
4. Soils supply many of the chemical elements required by living organisms.

Few soils totally fail physically to support some plants; however, some shallow soils are unable physically to support large trees (Figure 1–2). Each soil has a fixed location on the land and therefore is subjected to the climatic temperature and moisture dynamics of that location. However, specific locations with respect to the slope of the land and physical properties of the soil determine how water moves within the soil and significantly alters both the temperature and moisture dynamics of individual soils. The inorganic minerals in soil are the sole source of many elements required by both plants and animals. Although no soils are known to be totally devoid of the elements essential for life, there are vast differences in the quantities present and more significantly the rate at which growing plants can acquire life-essential elements from various soils.

Flowerpots or planters contain soil in a partial sense, but soil material removed from its spatial association with land is more nearly defined as dirt, that is, "loose or packed soil." In a similar fashion, soil samples taken to a laboratory for analysis are in effect not soil but rather soil material representing only a portion of the soil captured at a moment of time and not totally representative of the dynamic conditions soils experience in their natural setting within the land.

FIGURE 1–2   Shallow-rooted tree blown over by wind in the Appalachian Mountains of North Carolina, USA.

How many soils? Each person asked will answer differently. Those that have observed the color of soils may refer to red soils or black soils, reflecting what they have seen on road banks and in cultivated fields. Gardeners may respond by identifying sandy soil and clayey soil, reflecting their experience tilling the soil. Environmentalists may identify wetland soil or eroding soil. Bankers and farmers may refer to productive soils and poor soils. Many people have casually walked on soil and recall only muddy soil or dusty soil, depending on when they walked on the soil.

If a more detailed evaluation of soils is conducted, we find that within the United States the soil scientists name and classify about 22,000 different kinds of soil. Worldwide there are no estimates, but certainly the number is several times as many as recognized in the United States.[1] On the land, each kind of soil is laterally bounded by another kind of soil or, in some cases, barren rock, ice, or deep waters not capable of supporting rooted plants. The lateral boundary between different kinds of soils is often diffuse and not easily identified by observations of only the soil surface.

Different kinds of soil are determined by measurements of its ability to supply water to plants, temperature, mineral and chemical composition, and depth to material not capable of supporting plant roots. Whatever the number of soils in the world, each is well known and familiar to some people, but the vast majority of soils are unknown and never seen, much less experienced, by most people. Like other objects in nature, only samples of soils can be collected, transported, and displayed in a museum. An understanding of the dynamic interactions of soil and life-forms can be obtained only by nearly continuous observations of soils in their natural habitat within the land.

Life is characterized by the ability to perform metabolic functions of growth, reproduction, and some form of responsiveness or adaptability. Life-forms are composed of organic compounds of carbon, hydrogen, oxygen, and nitrogen obtained from air and water and other life-essential elements obtained from the inorganic minerals in the soil. Within plants these elements are formed into organic compounds such as sugars, starches, carbohydrates, and proteins. Animal life depends on the organic compounds formed by plant life for nutrition.

Upon the death of plants and animals, the organic compounds of which they are composed are decomposed by microbial life in and on the soil. As the organic components of life-forms decompose, the elements contained return to inorganic compounds in the air, water, and soil from which they can nourish another generation of plant and animal life on the land and in rivers, lakes, and oceans. Like plants and animals, microbial life-forms also grow by ingesting life-essential elements to form organic compounds within their cells, die, and are decomposed into inorganic components by succeeding generations of microbes. The decomposition of dead organisms is not an instantaneous process. Organic compounds of microbial life and dead organisms in various stages of decomposition are present within all soils and referred to as soil organic matter (SOM).

HUMAN FOOD AND THE LAND WHERE IT IS PRODUCED

Humans derive most of their food from only a very small and select group of plants and animals among the many species that naturally inhabit the lands of the world. Humans can consume most plant and animal tissue if proper preparation is provided to safeguard against poisons contained in some plants. Plant species that grow rapidly and produce quantities of proteins, carbohydrates, starches, and sugars in forms that can be stored for use as food throughout the year are favored. Rapid and reliable growth is desirable to enable the production of human food in areas where the growing season for plants is limited by cold conditions or seasonal rainfall patterns. Another property of a desirable food plant is a reproductive system that allows humans to easily establish a new generation of plants. Plants that reproduce via seeds that can be stored and transported are favored.

Seldom is an entire plant consumed for human food. Most plants concentrate life-essential elements in specific parts of the plant. The seed is often the most nutritious part of the plant for human food, although the stems, leaves, tubers, and fleshy rootstocks of some species are desirable human food. Cassava (Figure 1–3) is an example of a food plant that produces an eatable rootstock. Cassava is not a particularly fast-growing food crop, but in warm tropical climates it is a reliable source of food able to survive low soil fertility, drought, and many pest problems. Cassava and its many close relatives in the genus *Manihot* are as much a staple

FIGURE 1–3   Farmer in Brazil, SA, with a cassava plant. Cassava produces a fleshy, nutritious starchy rootstock and is cultivated as a staple food crop throughout the warm tropics.

human food in warm tropical lands as potatoes are in the colder areas of the world where cassava is little known except after it has been processed into tapioca. Corn is a cereal grass like wheat, oats, and several other cereal grains highly prized as food crops throughout much of the world. Although the natural origin of corn is debated, it was widely domesticated by indigenous peoples in the Americas from Chile to Canada at the time Europeans arrived in the New World. Corn and the other cereal grains have under gone many genetic changes at the hands of humans through selection of seed, hybridization, and recently developed techniques of genetic engineering (Prindle, 1994). Perhaps most people recognize varieties of corn developed for popcorn or sweet corn more readily than the grain corn that is processed into many of our common foods, cooking oil, and flour (Figure 1–4). Several cereal grains commonly grace our tables as major components of bread, cake, and numerous other foods, but strangely their plant name is most often preserved in breakfast products such as corn flakes, oatmeal, Wheaties, and Rice Krispies.

The number of plant species used for human food increases when animals are used to graze a wide range of plant species and produce eggs, milk, and meat for human

FIGURE 1–4 Ear of corn. Corn is a basic food crop from which many human foods are manufactured.

FIGURE 1–5 Dairy cows in Wisconsin, USA, on their way to graze grass on the hillside in the background that is not suitable for growing food crops.

consumption and/or fiber such as wool for human use (Figure 1–5). Although most animals can consume the same foods as humans, they are also able to consume plant species not palatable to humans. Animals are also mobile and able to access portions of the landscape not amenable to the human practices of growing food plants. Most animal flesh, milk, and eggs are palatable for human food, but many species of animals have defied domestication. Over many centuries, humans have domesticated and through selection of desirable characteristics genetically selected certain once wild animals into breeds that are most productive of human food and fiber.

Not all land is capable of producing human food. In 1997 the Food and Agriculture Organization of the United Nations (FAO) reported that only 11% of the land area in the world had no limitations for human food production (Table 1–1).

TABLE 1–1 Land Area and Soil Limitation for Agriculture

Percentages of Total World Land Area	Soil Limitation for Agriculture
28	Areas too dry
23	Chemical problems
22	Soil too shallow
10	Soil too wet
6	Soil too cold (permafrost)
11	No limitations

Source: FAO, 1997.

If we examine the limitations cited for the remaining 89% of the world, we quickly determine that it is probably not feasible to do much about heating those extremely cold areas with permanently frozen subsoil (permafrost). Also, little can be done to deepen soils that rest on hard rock and are too shallow to support the roots of food plants adequately. Although many areas that are naturally too wet for agriculture have been made productive by the installation of engineered drainage systems, this is an expensive practice and continued maintenance is required. Areas too dry for agriculture can be irrigated if there is a sufficient supply of salt-free water. Often these irrigated areas are very productive, but the supply of suitable irrigation water is limited and only a small fraction of land too dry for agriculture can be irrigated.

Most of the chemical problems that limit agriculture can be generalized as soils with natural chemical compositions inadequate for rapidly growing food crops. This does not indicate no vegetation will grow on these soils that often support lush forest growth. Distinct differences exist between the chemical requirements of natural vegetative ecosystems and crops grown for human food. Although the same life-essential elements are required for both natural ecosystems and human food crops, human food crops most often require much greater quantities of essential elements over shorter periods of time to produce satisfactory growth than most natural ecosystems. Since about 1850 the chemical composition of soil necessary for food crop production has been intensively studied and better understood by agricultural scientists. The information gained since then has resulted in the development and utilization of agricultural practices that chemically enrich and sustain the soil used for growing human food via chemical fertilization. Although chemical fertilizers are known to enhance food crop production on all land with amenable temperature and moisture conditions, the use of chemical fertilizers has not been uniformly applied throughout the world. The most extensive use of chemical fertilizers has been greatly to increase the productivity of food crops in those areas identified by the FAO as having no limitations for agriculture. The use of chemical fertilizer in conjunction with genetically improved seed and pest control on those lands has resulted in food crop yields per unit of land that are routinely more than double or triple crop yields per unit of land area achieved without chemical fertilization. The enhancement of food production via improved chemical fertility of the soil is presently not a viable option in many parts of the world because of a variety of political, economic, and social barriers.

OTHER HUMAN USES OF LAND

Several other human activities compete with food production for land area. Cities compete for land to be used for construction of shopping malls, parking lots, and housing developments. Often some of the best farmland is also the most desirable for these activities, and economic returns from urban development place a higher economic value on land than can be realized when that same land is used for food production. Social priorities for the preservation or restoration of natural ecosystems have resulted in abandonment of farming in some areas.

Societal desire for products other than food has prompted changes in the type of crops grown. Plants are composed of organic carbon-rich materials that can be converted into fuel. Following worldwide oil and gasoline shortages in the 1970s, in

FIGURE ... own in warm tropical ... into ethanol fuel for

Brazil, w...   public demand for gasoline pron...   ...ously used for food crops
(Figure 1...   ...ome cars and trucks were
modifiec...   ...added to almost all gasoline. Eth...   ...een added to gasoline in
other pa...

W...   ...d throughout the world.
Wood fo...   ...in many markets of the
world (F...   ...len tools such as the mortar and p...   ...or baking in wood-heated
clay ove...   ...y supply of fuel wood is a
necessar...   city or other fuels are not
availabl...   ...ere natural tree growth is
limited r...   ...arce, dried animal manure
and stra...

Co...   ...oducts, construction, and
veneer o...   ...o what humans have done
with foo...   ...than most food crops and
commer...   ...dom undertaken except in
econom...   ...g and harvesting valuable

FIGURE 1–7 Photo of a public market in Africa where among other produce, wood for cooking fuel is offered for sale. Wood is scarce in many places.

timber species such as mahogany suitable for veneer is an extensive practice in natural tropical forests (Figure 1–9).

The ability to concentrate the production of food and other materials to satisfy human needs is the essence of urban-based societies evolving from hunting and gathering societies. Whenever the product of a naturally occurring plant species has been found to be of benefit to human well-being, people have found ways to enhance the production of that product by concentrating the desirable plants onto small areas of land well suited to that plant's growth. In so doing they have reduced the amount of land needed for its production and reduced the need for human encroachment into natural ecosystems in search of the product.

WHO PRODUCES HUMAN FOOD?

The requirement for human food is fundamental for all societies. All too frequently the production of food is seen as the sole responsibility of the tiller of the land, that is, the farmer. Although the farmer has a primary role in human food production, the complexities of getting the food from the field where it is grown to the plates on our dinner tables involve the entirety of society. Farmers throughout the world have developed and utilize farming practices in response to both the physical and chemical constraints of the land they till *and* the economic, political, and social environment of the society in which they live. Collectively, the various systems of food production and distribution to

FIGURE 1–8   Photo of a typical indigenous family kitchen that includes a mortar and pestle for grinding grain into flour, a wood-fired oven, and a stove for cooking. Water is carried from the nearest stream, and there is no refrigeration.

human populations are best known as agricultural systems. Although the physical and chemical aspects of food production from the land (farming) are of primary concern, localized economic, political, and social conditions often determine many of the practices adopted by farmers around the world.

IF IT WORKS HERE, WILL IT WORK THERE?

In this book we explore the interactions of soil, land, and life from the perspective of human life and the role of human societies in the management of land for human benefit. Soils and land differ greatly in the type and severity of physical and chemical challenges they present to those who attempt to grow human food. Obviously, civilizations have found many ways to sustain food production for a long time. In various chapters, questions about management practices such as the following are discussed. Considering the composition and dynamics of different types of soil, what management techniques are most applicable? (Compatibility) How long will a management practice produce satisfactory results? (Sustainability) What are the social, political, and economic components of food production required for maintaining human food production? (Infrastructure) Will natural ecosystems be changed or destroyed in the quest to obtain human food? (Degradation and Pollution) If a soil fails human expectations, can it be altered by the application to technology and engineering? (Remediation) Can

FIGURE 1–9 Photo of logs for veneer harvested in the Amazon jungle of Peru, SA, being shipped on a riverboat. Seeking natural sources of high-quality hardwood trees that can be worth $100,000 US each often initiates road construction into remote jungles.

or should land not presently used for producing human food be brought into production with concomitant alteration of natural ecosystems? (Social and Political Priorities)

There is no one answer to any of these questions because neither soil, land, nor life is a singular, and societies differ in political, social, economic, and technical composition. Throughout the world someone or a group of people has experience obtaining food on each different soil and within a multitude of political, economic, and social conditions that have prevailed over time and presently exist in various countries of the world. An apparently logical solution to understanding the technical problems of managing different soils for food crop production would be to ask the local expert. Local experts are most often indigenous people tilling the land. Indigenous farmers are an invaluable source of knowledge; however, they have experience with only the traditional technologies and temporal economic conditions existing in the area (Figure 1–10). They know from experience what locally available management techniques are successful, but they are seldom prepared to explain the chemical, physical, and biological reasons for the success. They are unprepared to give enlightened understanding about the potential improvements or consequences of new technology or economic realities that they have not experienced.

FIGURE 1–10   Photo of an indigenous farmer, with his daughter, speaking to a group of international agricultural scientists in Peru, SA. It had not rained for several weeks, and the soil was dry and hard. When asked how he could plant in such hard soil, he immediately replied with an indigenous intellect that had escaped the questioner, "I wait until it rains."

The first objective of this book is to provide an overview of the physics and chemistry that bind the organic and inorganic entities of soil, land, and life necessary for human food production. The overview attempts to relate the interactions of mineral materials in soils, weather and climatic dynamics, and human intervention in altering these conditions to better serve human needs for food and fiber. Particular emphasis is given to outlining the fundamental physical and chemical requirements as life-forms interact with individual parcels of land and soil. A detailed presentation of the physical, chemical, and biological processes necessary for in-depth studies of soil science, geology, geography, and biology is avoided.

The second objective is to relate some insight into understanding the cultural activities that societies pursue in attempts to produce the necessities of human life in various parts of the world. In this regard an attempt is made in later chapters to identify the role of political, social, and economic infrastructure in determining why people have developed and presently use different agricultural strategies to cope with human food and fiber needs in various parts of the world.

LITERATURE CITED

Food and Agriculture Organization. 1997. Soil Limits Agriculture. Accessed at www.fao.org/NEWS/FACTFILE/FF9713-E.HTM

Lawdermilk, W. C. 1953. Conquest of the Land Through 7,000 years. Accessed at http://www.nativehabitat.org/conquest-1.html

Prindle, T. 1994. Native American History of Corn. Accessed at http://www.nativetech.org/cornhusk/cornhusk.html

NOTE

1. The appendix presents a brief description, distribution, and land use summary of the 12 orders of soils as classified by the National Cooperative Soil Survey of the United States.

CHAPTER REVIEW PROBLEMS

1. Of the 12 soil orders, which is most prevalent in the Midwest of the United States? In Eastern Europe? In Argentina?
2. Which of the 12 soil orders is composed of organic residue?
3. Which of the 12 soil orders has permanently frozen subsoil?
4. How many different soils are identified in the United States?
5. What is the difference between a soil and a soil sample?
6. What are the desirable characteristics of plants cultivated for human food?

CHAPTER

2

Water in Soil, Land, and Life

Water is necessary for life. In this chapter we trace the pathways of water as it falls from the air onto the land, passes through soil and plants, and returns to the air. Land and soil are fixed in place. Therefore the weather events that are summarized to define the climate of a location are a characteristic of each soil and parcel of land. Humans can do little to affect the amount, frequency, or rate at which water falls from the air as rain, sleet, or snow. Humans do manipulate the surface of the land in ways that influence the course water takes after it has fallen from the sky. Almost all human activities alter the characteristics of the soil surface. In urban areas the construction of buildings, roads, and parking lots eliminates the entry of water into the soil and forces rainwater to flow onto adjacent land. Planting, harvesting, and other cultivation activities routinely alter the condition of the soil surface several times each year where human food crops are grown.

THE RAINS COME AND GO

As we all know, rainfall is not continuous anywhere in the world. Actively growing plants require water every day. The universal role of land and soil is to harmonize sporadic rainfall events with the much more nearly constant physiological requirements growing plants have for water. The success to which water from natural rainfall is retained within the soil and available to plant roots between rainfall events in large measure determines the composition of natural ecosystems and the success of food crop production. Before water can be retained in soil, it first has to enter the soil. Human activities that alter the properties of the soil surface play a vital role in determining the course water will take on much of the earth's land surface. Water that falls as liquid or frozen precipitation eventually returns to the air as vapor to again condense and fall as precipitation. This cyclic movement of water is known as the hydrologic, or water, cycle (Figure 2–1).

A drop of water falling on the soil can follow several routes. If the water enters the soil, the process is called *infiltration*. If the rate of rainfall or snowmelt is greater than the rate of infiltration, additional rainwater most often flows over the soil surface as *runoff* water and becomes *run-on* water in adjacent land where it may infiltrate or continue to flow into rivers, lakes, streams, or oceans. The direct return of water in the form of vapor from a water surface to the air from is called *evaporation*. It occurs from all water surfaces

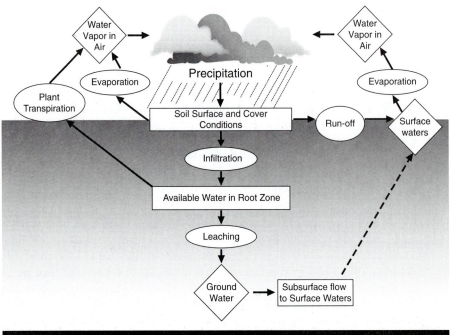

FIGURE 2–1 Schematic of water flow pathways in the hydrologic cycle.

in rivers, lakes, and oceans. Most water that temporarily remains on plant leaves or as puddles on level land surfaces after a rain event also evaporates to the air as water vapor. Most water that condenses on vegetation as dew also evaporates directly to the air.

Which path water from rain or snowmelt follows depends on the rate at which it arrives, the shape of the land, and the condition of the soil surface at the time of rainfall or snowmelt. On steeply sloping land, water is more likely to flow over the soil surface as *runoff* rather than infiltrating into the soil than on level land. The more level land adjacent and below the sloping land receives that runoff water as the speed of surface flow decreases. The "extra" *run-on* waters are likely to infiltrate the soil at those locations in the landscape.

The intensity of the rainfall greatly influences the relative proportions of infiltration and runoff. If the rate of rainfall is intense, a greater portion of the rainwater runs off than during a low-intensity rainfall. The relative proportions of runoff and infiltration changes as the surface conditions of the soil change from day to day or even minute to minute during a rainstorm. At the onset of a rain event, the soil surface may be very dry and most of the rainwater infiltrates. As the rain continues, the voids in the soil fill with water and more of the rainwater runs off.

Water that infiltrates into the soil is drawn downward by gravity and into smaller voids by capillary forces where some is stored as *available water in the root zone* for use by plants between rainfall events. If all the voids in the soil capable of storing water for plant use become filled with water, some water *leaches* below the rooting depth of the plants and eventually enters the *groundwater*. At some depth in or below the soil, all the soil voids are filled with water because there is an underlying layer that does not permit further downward movement of water. This zone is known as *groundwater*. The upper surface of the *groundwater* is known as the *water table*. Ground water slowly moves to low points above the underlying impermeable layer and eventually emerges, often at a great distance from the site where it infiltrated as surface water in a spring, river, lake, or ocean where it evaporates and returns to the air as water vapor.

A small amount of water, called capillary water moves into small voids above the water table. The rate and distance capillary water moves is variable depending on the size of the voids. The term is considered unconventional by soil scientists although in some situations it is an important process in creating saline conditions as discussed in chapter 6.

Water that enters the plant through the roots is vital for all physiological functions within the plant. Water escapes the plant as water vapor through openings on the green tissue of the plant and returns to the air by the process of *transpiration*.

SURFACE OF THE LAND-SOIL

Although the slope of the land and the intensity of rain event greatly influence runoff and infiltration, the conditions at the surface of the soil have a strong influence in determining whether rainwater infiltrates or runs off. The volume of all mineral soils is approximately 50% solid particles and 50% voids or pores formed between the solid mineral particles. The size and shape of voids in the soil are in large part determined by the size and shape of the solid particles. By convention, solid particles of soil are named by their diameters (Table 2–1).

The human eye without aid of magnification is capable of seeing objects as small as about 0.004 to 0.008 inches (0.1 to 0.2 mm) in diameter. Without a powerful microscope we can never see individual clay particles or even most silt-sized particles, much less the void space between these small particles. It takes 1,000 of the largest clay particles laid in a line to equal the diameter of the largest sand particle. Clay particles have a tremendous amount of surface area relative to their weight (Box 2-1). Clay particles are unseen even when using most optical microscopes that only allow us to see particles larger than about 0.0002 inches (0.005 mm) in diameter. Clay-sized particles can be observed only with electron microscopes.

With rare exception, all soil is a mixture of sand-, silt- and clay-sized particles. The relative proportions of these different particles result in the formation of soil textural names such as loam, silt loam, sandy clay, and so on (Figure 2–2). Particles larger than

TABLE 2–1 **Classification of Solid Particles in Soil by Particle Diameter**		
Name	*Millimeters*	*Inches*
Clay	< 0.002	< 0.00008
Silt	0.002–0.5	0.00008–0.0197
Sand	0.5–2	0.0197–0.079
Gravel	2–76.2	0.079–3
Cobble	76.2–254	3–10
Stone	> 254	< 10

Source: Soil Survey Staff, 1951.

BOX 2-1

Surface Area Related to Particle Size

Chemical reactions take place on the surfaces of solid particles in the soil. Smaller particles have more surface area per unit weight than larger particles. The importance of clay-sized particles relative to their weight can perhaps best be recognized by calculating the surface area of a solid sand-sized particle that is a 1-mm (0.0394-inch) cube weighing 2.6 mg (0.0026 g or 0.00000009152 ounce). The surface area of the six sides of that cube is 6 mm^2 (0.009 square inches). If we then divide the cube into eight equal cubes each 0.5 mm on a side, we have 12 mm^2 of surface area. The following table carries this division of each cube formed until the 1-mm sand size cube is the size of the largest clay particle and calculates the surface area in each step.

Number of Cubes	Cube Size	Surface Area	Particle Size Class
1	1 mm	6 mm^2	Sand
8	0.5 mm	12 mm^2	Sand
64	0.25 mm	24 mm^2	Silt
512	0.125 mm	48 mm^2	Silt
4,096	0.0625 mm	96 mm^2	Silt
32,768	31.25 μm*	192 mm^2	Silt
262,144	15.6 μm	384 mm^2	Silt
2,097,152	7.8 μm	768 mm^2	Silt
16,777,216	3.9 μm	$1,536 \text{ mm}^2$	Silt
134,217,728	1.95 μm	$3,072 \text{ mm}^2$	Clay

*1,000 μm equals 1 mm.

sand are not considered in the textural triangle, but if quantities of gravel, cobble, or stone exceed 15% of the soil, the textural name is modified as gravely-, cobbly-, or stony- (textural name), respectively.

Much attention is given to the analysis of particle-size distribution or texture of soil because many other features of a soil can be inferred from textural classification. Also, with practice, most people can very closely determine texture by feeling the soil. Perhaps the most important soil property that can be inferred from soil texture is the size of the voids formed between the particles as they pack together. Void size determines the ability of soil to transmit water during a rain event and retain water for plant use between rain events.

All soils have a complete range of void sizes. Sandy soils have a high proportion of large voids. Water infiltrates and flows rapidly through the large voids in sandy soils. A high content of clay in a soil indicates a high proportion of very small voids and slow infiltration and water movement. Soils containing high quantities of silt have more medium-sized voids than either sandy or clayey soils.

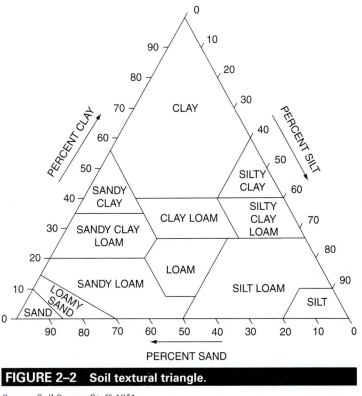

FIGURE 2–2 Soil textural triangle.

Source: Soil Survey Staff, 1951.

Void size in soil does not totally depend on texture. Insects, earthworms, and abandoned root channels also create large voids in the soil. When such voids intersect the soil surface, they are especially effective in increasing the rate of infiltration. The size of all voids in a soil is subject to change by activities that compress the soil. The largest voids, and especially those created by floral or fauna activities, are subject to closure by the physical impact of raindrops or compaction by animal or machine traffic on the soil surface.

RAINDROPS HITTING THE SOIL

Raindrops, especially large ones, falling from great height have a strong impact as they hit the soil surface. If large raindrops fall directly on the soil surface, they tend to compact the soil, close large void openings at the soil surface, and slow the rate of infiltration. In severe cases, such compacted layers of the soil surface become very hard as the soil dries and inhibit the growth of small seedlings. Compacted surface layers are known as *crusts,* and farmers may find it necessary to physically disrupt the crusts by shallowly tilling a soil that has been crusted by intense rain followed by rapid drying prior to seedling emergence. A microscopic view of a cross section of a soil surface that has been subjected to direct impact of raindrops reveals that no voids large enough to be seen with an optical microscope are present in a thin layer

Soil Surface

0.24 in.

FIGURE 2–3   Photomicrograph of a cross section through the *crusted* surface of a soil compacted by raindrop impact. Approximately 0.25 inch (6 mm) of the soil surface has few large voids, seen as black areas in the photomicrograph taken through a polarizing microscope. Such crusts reduce infiltration rate of water.

at the soil surface (Figure 2–3). Large visible voids are seen in the soil below the compacted surface.

If a soil is covered with dense vegetation it will more rapidly infiltrate water than the same soil without vegetation. In large part this is because the large raindrops first impact the vegetation where much of their impact energy is absorbed. As individual raindrops hit vegetation, they are broken into smaller drops that then fall to the soil surface from a lesser height and do not compact the soil surface or the water flows down the plant and has little impact on the soil surface. Vegetation, especially a dense grass, not only cushions the impact of the raindrops, but grass roots also create large vertical channels through the soil surface that provide pathways for water to infiltrate.

In the absence of cushioning vegetation, each drop of rain has the effect of a small bomb when it hits the soil surface. The larger the raindrop, the bigger the explosion. As large raindrops impact an unprotected soil surface, particles of soil are thrown up and in all directions away from the point of impact. On level land this causes compaction but because the particles are thrown equally in all directions, no net horizontal movement of soil occurs. On sloping land a net movement of splashed soil particles is down the slope in proportion to the percentage of slope. If the slope of the land is 10%,

FIGURE 2–4   Photo of small pedestals under areas where gravel protected the soil surface from the impact of raindrops that have eroded the surrounding soil materials during one rain event (US 25-cent coin for scale).

then 60% of the particles dislodged by the raindrop impact will fall down slope from the impact point. If some rainwater is not able to infiltrate, water will be moving down slope as surface runoff and carry the dislodged particles of soil further down slope.

Any object on or slightly above the soil surface that can absorb the impact of the raindrop, such as grass, leaves, or stones, protects the smaller soil particles from being dislodged and thrown into the air. Gravel-sized particles on the soil surface that are too heavy to be dislodged by the impact of raindrops often protect the soil under where they rest. In Figure 2–4 gravel is seen to rest on pedestals of soil more than an inch (2.5 cm) above the surrounding surface that has been eroded during one brief rainstorm. The same features can be seen at a much larger scale on barren land surfaces exposed to rain for many years (Figure 2–5).

TEMPORARY RETENTION OF WATER IN THE SOIL

The size and shape of the voids largely controls the movement and storage of water in the soil. After infiltrating into the soil, water is pulled downward by the force of gravity and pushed by subsequent water infiltrating the soil. Water rapidly moves downward under the force of gravity in large voids but is drawn into the smaller voids and retained, unable to move more deeply under the force of gravity by capillary forces of

FIGURE 2–5 Photo of a pedestal of soil protected by a petrified log in Arizona, USA. The surrounding area actively erodes in each rainstorm.

the smaller voids. Ability to supply the daily water needs of growing plants between rainfall events from water stored in these smaller voids is one of the soil's most vital functions. Before we look at the dynamics of retaining water in soil and examine how plants are able to satisfy their water requirements between rainfall events, we must define some standards for measuring water content in soil.

Water and air alternately occupy the void spaces between the solid particles in the soil. The size of the void determines under what conditions it will be filled or partially filled with water and when that water will be removed and replaced by air. Figure 2–6 shows void size and the relationship to the different states of soil water.

The content of water in soil greatly changes the weight of a given volume of soil. It is imperative that a common basis of water content and thus weight of a soil sample be used to compare all chemical and physical measurements among soils. To standardize soil weight for chemical analysis, soil samples are heated to between 221°F and 230°F (105°C and 110°C) to evaporate all the water. Heating to these temperatures does not burn the organic compounds present in the soil and creates a condition called *oven dryness* that represents a state where all voids are free of water and filled with air. This is the weight of the soil used as the basis for reporting most analytical data. The state of *oven dryness* rarely occurs naturally except occasionally in a very thin surface layer of soil not covered by vegetation.

When all the voids are filled with water and air is excluded, the soil is *saturated*. *Saturation* is present in the zone of groundwater. When a soil is saturated, water is

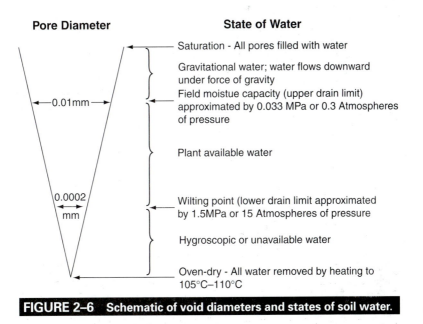

Pore Diameter   **State of Water**

Saturation - All pores filled with water

Gravitational water; water flows downward
under force of gravity

←0.01mm→

Field moistue capacity (upper drain limit)
approximated by 0.033 MPa or 0.3 Atmospheres
of pressure

Plant available water

0.0002
mm

Wilting point (lower drain limit approximated
by 1.5MPa or 15 Atmospheres of pressure

Hygroscopic or unavailable water

Oven-dry - All water removed by heating to
105°C–110°C

FIGURE 2–6   Schematic of void diameters and states of soil water.

available to plants but the roots of most crop plants suffer from a lack of oxygen
and/or an accumulation of carbon dioxide if saturation is continuously present more
than about three consecutive days. Some food crop plants, most notably rice, are able
to move oxygen from the leaves to their roots and thrive in saturated soil.

At moisture contents between oven dryness and saturation, soil voids are occupied by both air and water. The proportion of the void space occupied by air or water is
in an almost constant state of change except when the soil is saturated or oven dry.

When unsaturated soil is wetted by natural rain or irrigation, the water most
rapidly flows through the large voids. Water moving downward in the large voids in
response to gravity is known as *gravitational water*. As water moves downward in
response to gravity, it is drawn into the smallest voids by capillary force, progres-
sively filling larger voids as the smaller voids are filled with water. The distribution of
water in the various sized voids is not instantaneous. After a rain or irrigation event
stops adding water to the soil surface, the gravitational movement in the large voids
continues for about three days. After gravitational movement is too slow to measure,
the soil reaches a water content known as *field moisture capacity* (also called upper
drain limit). At field moisture capacity, all voids with diameters greater than about
0.01 mm are empty of water and those smaller are filled with water. In effect, the capillary forces of the void equal the force of gravity and downward water movement is
too slow for most practical considerations. A direct measurement of field moisture
capacity can be made by saturating a soil in place and then covering the soil surface
with impermeable plastic for three days to allow for gravitational flow. The soil is
then sampled. The moist soil sample is weighed and then dried to oven dryness and
again weighed. The difference in weight is the amount of water retained at field
moisture capacity. This is a time-consuming process, and field moisture capacity is
often approximated in the laboratory by subjecting samples to a positive air pressure

of 4.85 lb in^{-2} (0.033 megapascals [MPa] or 0.33 atmospheres [atm]). Some researchers consider a pressure of 1.47 lb in^{-2} (0.01 MPa or 0.1 atm) as a more appropriate approximation field moisture capacity.

Although plant roots are able to extract water from the larger voids, as they empty under the force of gravity they are able to exert forces greater than gravity and continue to extract some water held in voids smaller then 0.01 mm (0.00039 inches) in diameter. Plant species vary greatly in the amount of force they can exert to extract water held by capillary forces within the smaller voids, but most crops extract water from voids no smaller than about 0.0002 mm (0.0000079 inches) in diameter. The water content of the soil when plants visibly wilt and die from lack of water is appropriately called the *permanent wilting point* (also known as lower drain limit). The permanent wilting point can be directly measured by growing plants in a container of soil until they wilt and fail to recover overnight. At that point the soil in the container is weighed, dried to oven dryness, and weighed again to determine how much water was present at the time the plant died. Air in the larger pores is in direct contact with enough liquid water in the smaller pores that the relative humidity of air in the soil is 100% and plants do not instantly die when the permanent wilting point is reached. Several days of observing the plant wilt during the day and recover overnight may be required to determine the wilting point directly. Because this is a troublesome and time-consuming procedure, permanent wilting point is approximated by subjecting soil samples to 220.5 lb in^{-2} (1.5 MPa or 15 atm) of air pressure in an especially designed pressure apparatus to extract water from voids greater than 0.0002 mm in diameter. Water held in voids with diameters less than 0.0002 mm is not available to most plants and can only be removed by evaporation. This water has been termed *hygroscopic* or *unavailable soil water.*

The amount of water held in voids with diameters between 0.01 mm and 0.0002 mm (0.00039 inches and 0.0000079 inches) is available for plant use between rainfall events. The amount of water a soil can retain against the force of gravity but have available for plant use between rain events is called the *available water-holding capacity* of the soil.

Available water-holding content is often determined and expressed on a soil weight basis, but plants extend their roots into a volume or depth of soil and do not weigh the soil. Soils differ greatly in weight of soil per unit volume of soil (bulk density[1]) (Box 2-2). Expressions of available water-holding capacity on a weight basis accurately reflect the amount of water held in a volume or depth of soil only when the bulk density of the soil is 1 gram per cubic centimeter (0.036 lb in^{-3}). Weight-based values of water content underestimate volumetric water content in soils with high bulk density and overestimate volumetric water content in soils with low bulk density. Soils with high clay contents have lower bulk densities than sandy textured soils. Organic soils (Histosols; see appendix) have extremely low bulk densities. To obtain a volume of water per volume of soil, the weight percentage of water is multiplied by the bulk density of the soil and expressed as inches of water per inch or centimeters per centimeters of soil depth (Box 2-3). For example, the weight of available water (field capacity–permanent wilting) is 0.1 g per 1 g of soil (10%) on a weight basis. If the bulk density of the soil is 1.3 g per cubic cm 0.1 × 1.3 (bulk density of the soil) equals 13% water on a volume basis. Because the density of water is 1 g cm^{-3}, that equals 0.13 cm water cm^{-3} of soil or 1 cm of soil depth. If the same amount of water on a weight basis

BOX 2-2

Specific Gravity, Density of Soil Particles, and Soil Bulk Density

The ratio of the density of a substance to the density of another substance is a unit-less value called specific gravity. Most minerals in soil have a specific gravity of 2.65 when immersed in water. Water (at 4°C) has a defined density of 1 g cm^{-3}. Thus

$$2.65 \times 1\,g\,cm^{-3} = 2.65\,g\,cm^{-3} = 2.65\,Mg\,m^{-3}$$

(Megagram per cubic meter (Mg m^{-3}) is the preferred SI unit. (See Appendix II.)

The average density of mineral particles in soil is usually estimated as 2.65 Mg m^{-3}. However, minerals containing iron and other heavy elements have a higher specific gravity, and most clay minerals have a somewhat lower specific gravity. If mineral particles with a density of 2.65 Mg m^{-3} are packed leaving a 50% void space between the particles, the density of the resulting volume is 1.325 Mg m^{-3} and called *bulk density*. The bulk density values we actually find in soils differ greatly. Lower bulk densities are present when large quantities of organic matter are present in soil. Clay textured soils have a lower bulk density. Higher bulk densities occur when a soil is compacted and the void space reduced to less than 50%. Compact soil can restrict root growth. Although the ability of plant roots to penetrate compacted soil differs among plants, a bulk density greater than 1.6 Mg m^{-3} in sandy textured soil and 1.4 Mg m^{-3} in clay textured soil often restricts root growth.

Engineers most often use English units of lb ft^{-3} or lb yd^{-3}. (See Appendix II for conversions.)

BOX 2-3

Converting Weight-Based Measurements to a Volume or Soil Depth Basis

Oven-dry weight of soil is the basis for expressing all laboratory measurements of soil material. However, plants do not weigh the soil they root into. Rather, plants root into a volume of soil best identified as a depth of soil. The soil measurement used to convert any value measured on the basis of soil oven-dry weight to a volume or depth of soil is *bulk density* (BD), the weight of a unit volume of soil expressed by this formula:

BD = weight of soil in grams/volume of soil in cm^3, i.e., g cm^{-3} or the preferred SI expression:

BD = weight of soil in megagrams (Mg)/1 cubic meter volume, i.e., Mg m^{-3}.

(Note: g cm^{-3} and Mg m^{-3} have the same numerical value.)

Scientists almost always use the metric system because 1 g of water occupies 1 cm^3, making for easier conversions. Engineers frequently use pounds per cubic foot or cubic yard.

To determine the bulk density of soil, a known volume of soil is extracted from the soil, usually by carefully inserting a small thin-wall metal cylinder of known dimensions

(continued)

(*Continued*)

into the soil so as not to compress it. After removing the soil-filled cylinder and carefully scraping off any soil that protrudes, all of the remaining soil is removed from the cylinder, weighed, and then dried at a temperature of 221°F to 230°F (105°C to 110°C) and weighed again. If, for example, the soil sample from the cylinder weighs 161 g before it is oven dried and weights 140 g after it is oven dried, it contains 21 g of water. That amount of water would be expressed on a weight basis as follows:

$$(21 \text{ g water})/140 \text{ g oven-dry soil} = 0.15 \text{ g water/g soil or 15\% by weight}$$

If the volume of cylinder is 125 cm^3, the bulk density of the soil is:

$$BD = 140 \text{ g oven-dry soil}/125 \text{ cm}^3 = 1.12 \text{ g/cm}^3$$

Because it was determined that a gram of soil contained 0.15 g of water, the formula to convert to a volume basis would be:

$$(0.15 \text{ g water/g soil}) \times (1.12 \text{ g soil/cm}^3 \text{ soil}) = 0.168 \text{ g water/cm}^3 \text{ soil}$$

Because the volume of 1 g of water is 1 cm^3, the equation could be written:

$$(0.15 \text{ cm}^3 \text{ water/g soil}) \times (1.12 \text{ g soil/cm}^3 \text{ soil}) = 0.168 \text{ cm}^3 \text{ water/cm}^3 \text{ soil}$$

Because 1 cm^3 is also 1 cm thick, both the numerator and denominator can be divided by cm^2 to obtain:

$$0.168 \text{ cm of water/cm of soil depth}$$

Because both the water and soil depth units are in centimeters, both can be multiplied by 2.54 cm/inch to arrive at:

$$0.168 \text{ in of water per inch of soil depth}$$

Soils differ greatly in bulk density. The differences are mainly related to texture, degree of compaction, and organic matter content. When the quantity of any soil component is compared only on an oven dried soil weight basis, that quantity is overrepresented on a soil depth basis in soils that have high organic matter or clay contents and a low bulk density and underrepresented for soils with a high bulk density. Simply stated, to get the same amount of water or other material expressed as equal on a weight basis, plant roots need to search a greater volume of low bulk density soil. Weight-based values equal volume-based values only when the bulk density of the soil is 1 g cm^{-3}.

The following table contains comparisons of equal per weight values and per depth values when bulk density values are considered:

From the table it should be clear that a plant would have to root more than twice as deep in an organic soil than in a silt loam soil to obtain the same amount of water or any other material that was expressed as equal on a weight (g g^{-1}) basis.

Soil	Bulk Density g cm^{-3}	Weight per Unit Weight g g^{-1}	Weight per Unit Soil Depth g cm^{-1}
Sand	1.6	0.2	0.32
Silt loam	1.3	0.2	0.26
Clay	1.1	0.2	0.22
Organic soil	0.6	0.2	0.12
Compact clay	1.5	0.2	0.30

TABLE 2–2 Approximate Available Water-Holding Capacity Related to Soil Texture

Texture	Amount of Available Water (Either as inches of H_2O per inch of soil depth or cm of H_2O per cm of soil depth.)
Sand	0.05
Loamy sand	0.1
Silt loam	0.2
Clay	0.15

is present in an organic soil with a bulk density of 0.5 g cm^{-3}, the amount of available water on a volume basis is only 0.05 cm in 1 cm depth of soil, that is, $0.1 \times 0.5 = 0.05$. Both the volume of water and soil depth are in the same units and can be expressed as inches of water per inch of soil depth (i.e., 0.13 inches and 0.05 inches of available water per inch of soil depth in the preceding examples).

Available water content per unit depth of soil is germane to understanding the interaction of plants and soil water. The available water-holding capacity of most soils range between 0.05 inches of water per inch of soil depth and 0.2 inches of water per inch (0.05 and 0.2 cm per centimeter) of soil depth. Knowing this value for any given soil, and the rooting depth of the plants growing on that soil, we can calculate how much water will be available for plant growth following a rain or irrigation event.

In a very general way, void size in soils is directly related to solid particle size. Sandy soils have a high proportion of large voids in the *gravitational* soil water range. Clayey soils have a high proportion of very small voids in the *unavailable* soil water range. Soils with a high proportion of silt-sized particles have a high proportion of pores in the *available* soil water range. Silt loam textured soils usually retain more available water per unit depth than either sandy or clayey soils and thus provide water for plant growth for longer periods of time between rainfall or irrigation events (Table 2–2).

LEACHING AND GROUNDWATER

When plants are actively growing, most rain events only supply enough water to partially fill all the voids smaller than 0.01 mm and replenish available water within the root zone in the soil. When rain or irrigation events yield a large amount of water, the infiltrated water rapidly moves downward through large continuous pores such as root channels or wormholes. If the rate of water infiltration is more rapid than movement of water into the smaller voids adjacent to the large pores and the large continuous channels are sufficiently deep, some water may not fully wet the smaller voids in the soil through which it passes. This rapid flow downward through large continuous pores is known as *bypass flow* and does not fully recharge the available water-holding capacity within the root zone. The amount of infiltrated water that is not retained by the small- and medium-sized pores or bypasses these pores moves to a depth not reached by plant roots and is lost (*leaches*) from the soil (Figure 2–1). Clearly this depth depends on the rooting depth of the species of plant growing and also changes with the stage of plant growth.

At some depth under most land there is a layer of *groundwater* in which all the voids are filled with water and a permanent condition of saturation exists. The depth of the upper surface of the *groundwater,* that is, the *water table,* varies greatly under land surfaces. In arid climates it may be several hundred feet below the land surface. In humid climates it is usually much less deep and tends to follow the contours of the landscape, being deeper under the hills and nearer the land surface in the valleys, as discussed in later chapters.

The lower boundary of groundwater is an impervious or slowly permeable layer of rock or sediment. Groundwater slowly flows from higher to lower elevations as defined by the elevation of the impervious or slowly permeable layer. At some point in most landscapes groundwater flows to the land surface as springs or enters rivers and lakes under the water level. The movement of water within the zone of groundwater is through voids in soil-like material or cracks and channels in the bedrock. By digging or drilling a well that extends into the zone of groundwater, humans can pump water for drinking or irrigation. At the location of a well the water table is lowered as water is extracted from the well. The localized depression of the water table is known as a zone of depression and after extraction of water ceases will slowly refill as water moves into the zone. A reliable well must extend deep enough into the groundwater that water recharges the zone of depression at a rate fast enough to satisfy the extraction rate of water from the well. Depth to the water table is directly related to the cost of pumping water to the surface.

Water that infiltrates soil and leaches to the groundwater can either become contaminated by dissolving chemicals present in the soil through which it passes or cleansed by the filtering action of the soil and geologic material through which it passes above and in the zone of groundwater. Most groundwater is usable for human consumption. However, groundwaters differ in taste and chemical composition, largely determined by the composition of the material through which it passes. Local concentrations of both natural- and human-induced undesirable substances sometimes find their way into groundwater, rendering it unusable for human consumption.

ALTERATION OF VOIDS IN THE SOIL

Sandy textured soils with a high proportion of large voids have a high infiltration rate and drain rapidly after a rain. Clay textured soils have a slow infiltration rate and remain near saturation for a longer period of time as water moves more slowly in response to gravity. However, these generalizations are subject to extreme variation because soils are not uniform mixtures of particles. Large voids may be present in soils because decomposed roots leave open channels. Earthworm holes are also an example of this kind of alteration. The large continuous channels left as plant roots die and as wormholes are temporary features subject to closure by as little as the footstep of a deer. Thus the same soil when permeated with roots and earthworms may have considerably different water movement characteristics than when it is not vegetated or compacted by vehicle or animal traffic.

Many soils undergo freezing and thawing cycles one or more times during the course of each year. Voids filled with water become plugged as ice forms. Many soils in polar latitudes have permanently frozen subsoil (see Gelisols in the appendix) that stops the downward movement of water, causing the soil above to be saturated as the surface layers of soil warm and melt each summer. In the colder parts of temperate latitudes, soils are often frozen each winter. Each spring as the soils thaws from the

0.1 mm

FIGURE 2–7 Silt-sized particle of mica. Such flat particles often become oriented parallel to the soil surface by raindrop impact and contribute to reducing infiltration.

surface downward, temporary saturated conditions are present near the surface. Water from melting snow and all rainfall that occurs during that time runs off the soil surface, making springtime a prime time of flooding along the rivers.

Management does not change soil texture. However, the size and shape of voids can be altered by cultivation, animal or insect activity, and plant roots. Not all particles in the soil are spherical. In fact, most clay and many silt-sized and some sand-sized particles are platelike in shape (Figure 2–7). When wet soil is subjected to compaction from animals or vehicles, these platelike particles tend to orient themselves parallel to the soil surface like shingles on a roof and slow downward water movement. Compacting action that closes large voids and channels is said to *puddle* the soil, perhaps because infiltration is slowed and puddles of water often form on the soil surface after a rain event.

Farmers almost universally have found that it is desirable to disturb the soil surface prior to planting. These cultivation practices break up thin surface compaction that may have taken place from rain or compacting traffic prior to planting (Figure 2–3). Most cultivation practices in preparation for planting also create micro hills and depressions into which rainwater can collect and infiltrate. The moisture content of the soil at the time it is plowed or otherwise cultivated with tillage implements is critical. If soil is cultivated when it is too wet, compaction and orientation of platelike particles can take place both within and immediately below the layer tilled. Compaction collapses some of the large voids, thereby slowing the movement of water in the soil. When compaction occurs at the bottom of the tillage layer, it is called a *plow* or *tillage pan* and can restrict root growth and downward water movement, greatly reducing crop yield. Plow pan conditions are not usually permanent, and within a few years as wetting and drying cause slight volume changes, insects and earthworm activity create channels, and roots slowly penetrate the compacted layer and the pan restriction is removed. Pulling a deep chisel implement can disrupt most plow pans, but this requires considerable power and is expensive. Most farmers avoid creating pans by being aware of the moisture content of the soil at the time of cultivation.

Tillage pan formation can be a problem for most crops, but it is highly desirable in flooded or paddy rice growing. Rice farmers often intentionally create a pan in their fields by mechanical traffic or having cattle or people walk on wet soil. The presence of water-restricting pans reduces the leaching loss of water from flooded rice fields.

WATER IN THE PLANT

Actively growing plants remove water from the soil and move it through their xylem cells to their leaves where it evaporates into the air from small openings called *stomates* (plural, *stomata*). This process is called *transpiration* (Figure 2–1). Transpiration begins each day as the stomata open to admit air and obtain carbon dioxide for the process of photosynthesis. The stomata close at night and transpiration ceases. A precise calculation of moisture requirements by a grain crop under field conditions is extremely complicated. Atmospheric variables that affect transpiration are air temperature, hours of sunlight per day, relative humidity, and wind velocity.

Air temperature has the greatest influence on the moisture requirements of plants and evaporation from the soil surface. Most studies combine transpiration and evaporation from the soil surface into the term evapotranspiration. Table 2–3 gives the approximate potential evapotranspiration rates for 14 hours of sunlight during summer days in temperate latitudes and daily air temperatures. In tropical latitudes with only 12 hours of sunlight daily, evapotranspiration rates are somewhat less for corresponding air temperatures. If there is a complete cover of green, actively growing vegetation present, the rate of water required for evapotranspiration is nearly the same for all types of vegetation. Individual trees not surrounded by trees of equal height will exceed the values in Table 2–3 because all sides of the tree are exposed. High wind velocity and low relative humidity increase transpiration rates.

TABLE 2–3	Approximate Potential Evapotranspiration Rate by a Complete Cover of Actively Growing Vegetation as Related to Average Daily Air Temperature*		
Average Daily Air Temperature		*Potential Evapotranspiration Rate*	
°F	*°C*	*Inches H₂O/Day*	*Centimeters H₂O/Day*
90	32.2	0.26	0.66
80	26.7	0.21	0.53
70	21.1	0.16	0.41
60	15.6	0.13	0.33
50	10.0	0.10	0.25

*Approximately 14 hours of sunlight per day.

Plants are able to reduce transpiration for several hours or even days by temporarily wilting during sunlight hours. Wilting is the outward sign that the stomata in the leaves are partially closed and the plant is not growing at a maximum rate. How severely wilting reduces the rate of plant growth and yield of food grain differs depending on stage of growth and crop species. Corn yields are severely reduced if inadequate water is available at the time of pollination. Estimates of potential evapotranspiration rates differ from actual evapotranspiration rates for several reasons. The most obvious reason is that there may be no actively growing plants as during winter months in many areas or when all vegetation has been removed by cultivation. When only small crop seedlings are growing and they do not completely cover the land surface actual evapotranspiration rates are less than potential evapotranspiration rates. Also, as crop plants near maturity and their leave are no longer actively growing the actual evapotranspiration rate slows.

WATER DYNAMICS DURING A GROWING SEASON

All soils have a pattern of wetting and drying in response to weather conditions and their position on the landscape. For the purpose of illustrating seasonal moisture dynamics in temperate latitudes as related to crop requirements, we can assume that a field of corn requires the following:

1. 0.1 inches (0.25 cm) of water per day for the first 25 days of the growing period when temperatures are rather cool, the plants are small, and leaf cover is not total.
2. 0.2 inches (0.5 cm) of water per day for 50 days in the middle of the growing season when the plants are actively growing and corn leaves completely shade the soil surface.
3. 0.1 inches (0.25 cm) of water per day during the last 25 days of the growing period when the corn plants are no longer actively growing and the seed grain is maturing.

For such a crop, a total of 15 inches (38.1 cm) of water would be required. Average rainfall expected during 100 days of the growing season at Peoria, Illinois, in the heart of the U.S. corn belt, is reported as about 12 inches (30.5 cm), leaving, on average, a shortage of 3 inches (7.6 cm) of water. Clearly, this area would not have attained a reputation as an excellent corn-growing area if it depended only on average rainfall during the growing season. From where does a crop obtain the extra water? In the example used, we must consider that at the time of planting, as temperatures warm in the spring season, the soil is fully charged with available water (all voids smaller than 0.01 mm are filled with water) from the winter period when there has been little or no transpiration. The water content in the soil at planting can be assumed to be at field capacity. With silt loam texture, each inch (2.5 cm) of soil depth would be expected to contain 0.2 inches (0.5 cm) of available water (Table 2–2). Thus, if the corn plants extended their roots to a depth of 15 inches (38 cm), they would access 3 inches (7.6 cm) of available water stored from the previous winter. Most soils in the Peoria, Illinois, area are friable and deep such that corn roots have been observed easily to extend much deeper than 15 inches (38 cm) and are therefore capable of accessing even more water should the rainfall during the growing season be below average.

To understand more fully the uncertain nature of moisture supply, let's assume there are 20 consecutive days during the most active growing period when no rain falls. During such a drought, the corn requires 0.2 inches (0.5 cm) of water per day to attain its full growth rate potential. This would be a total of 4 inches (10 cm) of water and would require the roots to access a 20-inch (50-cm) depth of soil that was at field moisture capacity prior to the onset of the drought. This is quite realistic for most of the silt loam textured soils in the Peoria, Illinois, area.

WATER PATTERNS ON THE LAND

It is not correct to assume that the scenario just described would be correct for all soils in the Peoria area. Some areas would undoubtedly have wilting and slower growth with reduced yields resulting from one or more of the following conditions. Sloping areas may not be able to infiltrate all the rainwater at the rate it falls during the intense thunderstorms of the summer season. For every inch (2.5 cm) of rainfall that runs off the surface, the plant roots would have to grow 5 inches (12.5 cm) deeper to access stored available water.

In some soils, a subsoil layer may restrict rooting depth to a lesser depth than required to access the necessary water. Also, available water-holding capacity of 0.2 inches per inch (0.5 cm per 2.5 cm) of soil depth is correct only for soils with silt loam textures. If the soil has a sandy texture and holds only about 0.05 inches (0.127 cm) of available water per inch (2.5 cm) of soil depth, an 80-inch (203-cm) rooting depth would be needed to supply 4 inches (10 cm) of water during a 20-day drought.

Perhaps the most visible pattern of spatial variations in moisture supply can be observed in the growth of crops in rolling or hilly landscapes. If rainwater runs off the surface of a hillside, much of that water infiltrates into the soils at the base of the slope. Thus, in seasons of extreme draught, when whatever rain does fall comes as intense thunderstorms, an extreme reduction in crop growth will be visible on the hillsides while the level land at the base of the slope receiving run-on water often has excellent plant growth (Figure 2–8).

The same area, in another year with above average rainfall may have quite contrasting results. The same process of runoff from the hillsides may, in some years, result in too much run-on water in the depressions. This may delay planting or in extreme seasons saturate the soils at the base of the slope for long periods, and the roots will suffer from lack of oxygen or high carbon dioxide levels, at which time they are very subject to disease. This may result in a crop failure at the base of slope while the adjacent hill slopes may have a good crop that growing season.

An indication of how soil texture and landscape differences can affect crop production is illustrated by the expected long-term average corn yields published in the soil survey of Peoria County, Illinois (Walker, 1992). Upland areas with deep silt loam textured soils and slopes of less than 5% can expect to average 153 bushels of corn per acre (9,180 kg ha^{-1}). Similar soils on slopes from 5% to 10% can expect to average only 146 bushels of corn per acre (8,760 kg ha^{-1}). Landscape depressions that frequently become saturated are expected to average only 113 bushels of corn per acre (6,780 kg ha^{-1}). Deep sandy textured soils on 3% to 7% slopes are reported to average only 45 bushels of corn per acre (2,700 kg ha^{-1}). All the yields just cited are the

FIGURE 2–8   Photo of corn in Wisconsin, USA. Note the good growth of corn in the foreground at the base of the slope and the poor, almost absence of corn on the adjacent side slope where much of the water runs off leaving the soil too dry for good corn growth.

expected averages when competent farmers use accepted management practices of fertilization, cultivation, and cultivar selection. Yields in individual years may be significantly above or below these expected averages.

There is one critical feature in the Peoria, Illinois, scenario that is so common to temperate latitude inhabitants that it is frequently overlooked in tropical areas. In temperate latitudes the time of grain planting is dictated by warming temperatures in the spring of the year. This is a time when the soils contain all the water they can hold at field moisture capacity because of winter precipitation and lack of evaporation and transpiration during the cold weather. In much of the tropics, planting time has to correspond to the onset of a rainy season that often follows several months of rainless warm weather. At the onset of seasonal rains in the tropics, the soil contains almost no available water. With no supply of available water in the soil, the crop is totally at the mercy of the whims of precipitation. Should rains not be forthcoming at frequent intervals to supply water needed by the crop after planting, there is no reserve of water in the soil and a complete crop failure may result. Under such conditions a more reliable rainfall pattern during the growing season is necessary to avoid drought stress than in the Peoria, Illinois, scenario. This is often aggravated by the fact that few tropical areas have high silt-content soils but rather have soils with a much lower available water-holding capacity than those in the Midwest of the United States or northern Europe (Figure 2–9).

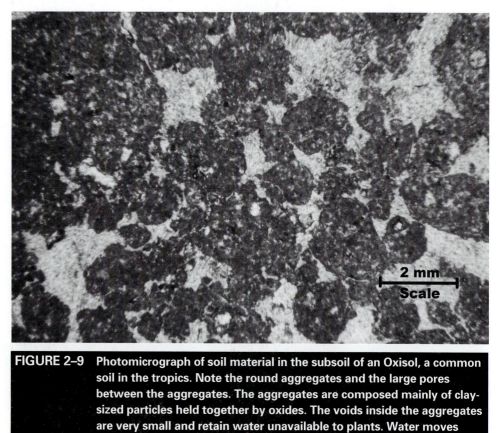

2 mm
Scale

FIGURE 2–9 Photomicrograph of soil material in the subsoil of an Oxisol, a common soil in the tropics. Note the round aggregates and the large pores between the aggregates. The aggregates are composed mainly of clay-sized particles held together by oxides. The voids inside the aggregates are very small and retain water unavailable to plants. Water moves rapidly through the large pores, often *bypassing* the clay aggregates. There are few pores of the size to retain plant available water.

WHEN DOES THE SEPTIC SYSTEM FAIL AND THE WELL GO DRY?

The answer to both of the questions posed in the heading to this section relates to the depth of the *water table* at a specific location. A water table is present at some depth under most land but more often near enough to affect septic system and shallow wells in more humid areas of the world. Depth to the water table changes from season to season depending on the amount of water that leaches and the rate at which the groundwater flows to springs, rivers, lakes, and streams. The rate of groundwater movement is slow because it is passing through relatively small pores in the soil or geologic material. During periods of excess rainfall and infiltration, the depth of the water table rises at a given location, and during drought it deepens. Wells depend on a saturated condition and fail if the water table drops below the depth of the well (Box 2-4).

Septic systems depend on unsaturated voids into which they can empty their watery effluent. If the water table rises above the septic system, the effluent escapes through the surface of the soil, often spreading across the front lawn (Figure 2–10).

BOX 2-4

Schematic of Septic System and Well Relationship to Seasonal Water Table Depth

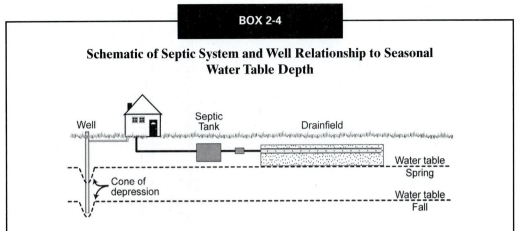

A septic system consists of a septic tank into which sewage effluent from the house empties. In the septic tank, microbes decompose most of the solid organic materials. Some solid materials are certain to remain, and it is necessary to have the tank cleaned out regularly. However, an all too common failure of the system lies not in the septic tank but in the drain field. As the septic tank fills, watery effluent from the top of tank flows into a series of porous pipes buried in trenches below the soil surface and downhill from the septic tank. From the drain field the effluent moves into "unsaturated" soil were it is further filtered and either leaches or is taken up by growing plants. As the water table rises, as it invariably does each winter and early spring in humid temperate latitudes to a level where the effluent cannot leach and vegetation is not actively growing, the effluent is forced to the soil surface (see Figure 2–10).

A well is simply an open pipe inserted into groundwater from which water is pumped. As water is pumped from a well more rapidly than groundwater can flow to the well, the water table around the well drops in an area known as the cone of depression. As long as the water table is shallow enough that water can flow to the well at a rate equal to pumping, the bottom of the cone of depression does not extend below the depth of the well and there is a constant supply of water. If during a hot, dry summer the water table drops to a level such that the pumping rate exceeds the rate at which water can flow into the well and the cone of depression deepens below the well depth, no water can be extracted. If pumping ceases, the cone of depression will slowly fill and some water can again be pumped at a slower rate.

In temperate latitudes and in locations where the water table is close to the land surface, the depth to the water table increases during the summer when plants are actively transpiring, and the septic system is not likely to fail but the well may go dry. During the winter, with little or no transpiration, the water table rises and the septic system may fail but the well has an abundant supply of water. In more arid areas the water table may never be close enough to the surface to affect a septic system. Deep municipal wells may experience a deepening of the water table resulting from several years of below normal leaching in the distant watershed that is the source of the groundwater they extract.

FIGURE 2–10 Photo of an urban lawn with effluent from a failing septic system flowing to the surface in North Carolina, USA. During the early spring season, the water table rises and effluent from the septic system flows to the soil surface, warming and fertilizing the lawn. Untreated septic effluent on the surface is a potential health hazard.

PERSPECTIVE ON WATER

Water is an imperative of life. Plants are able to survive on less than ideal amounts of water for various periods of time by wilting, but deprived of adequate amounts of water their growth is retarded. The total amount of water available is closely related to the geographic location of the land in relation to climatic conditions of both rainfall and temperature. Slope of the land and the condition of soil surface are major determinants of relative rates of water infiltration into the soil and runoff over the soil surface that result in a spatial distribution of water within local landscapes. The ability of soil to store and release water to plants between rain events is related to the volume of pores in the soil of a size that can retain plant-available water. The distribution of pore size in the soil is largely determined by the texture of the soil. These relationships are applicable to all areas of the world. Chapter 3 examines more fully how the interactions of water and temperature influence human activities on land in various parts of the world.

LITERATURE CITED

Soil Survey Staff. 1951. *Soil Survey Manual.* USDA Handbook No. 18. Washington, DC: U.S. Government Printing Office.

Walker, M. B. 1992. *Soil Survey of Peoria County, Illinois.* USDA Soil Conservation Service. Washington, DC: U.S. Government Printing Office.

NOTE

1. Bulk density is almost always expressed in metric units of g cm^{-3} (1 g of water has a volume of 1 cm^3).

CHAPTER REVIEW PROBLEMS

1. If the evapotranspiration rate is 0.2 inches of water per day, how deep must the plant roots extend to provide the water needed for 10 rainless days in a soil that retains 0.1 inch of available water per inch of soil depth?

2. What percentage of the volume of most soil is void or pore space?

3. Compare and contrast the processes of evaporation and transpiration.

4. Discuss how the rate of infiltration is related to the rate of runoff during an individual rain event.

5. Where in the soil is the unavailable water?

6. How much of pore space in the soil is filled with water at saturation?

CHAPTER

3

Temperature in Soil, Land, and Life

Every reaction that takes place in soil and life must obey the laws of chemistry and physics, and the speed of reactions is often related to temperature. Warm-blooded animals, such as humans, are able to partially control their body temperature. Plants and cold-blooded animals, which include all microorganisms and most animals and insects that live in the soil, cannot effectively control their temperature. They depend on the ambient soil temperature to determine if they grow rapidly or slowly, live or die. Both the temperature of the air and the temperature of the soil affect plants that exist partially in the air and partially in the soil. The temperature dynamics experienced by plants, cold-blooded animals, and microbes are primarily determined by the location of the land they inhabit.

The rate of biological reactions in all life-forms is closely related to temperature. On hot days plants transpire more water than on cold days (see Table 2–3). If the temperature falls below freezing, the water in some plant tissues turns to ice and that tissue dies. The actual freezing point of most plant tissue is usually somewhat below that of pure water because of dissolved chemicals in the cell water, and some plants are not damaged by extreme cold. Not all the parts of the same plant are equally susceptible to death by freezing. The early flowering stages of fruit trees and berries are usually most susceptible to frost damage. Freezing of the flower buds may completely eliminate the production of fruit that year while the stems and roots of the plants remain alive.

Of more practical concern to the life functions of plants and microbes is *biological zero,* the temperature at which the physiological processes in the plants and microbes become so slow that they essentially cease. For most organisms, biological zero is 39.2°F (40°C). Individual organisms and tissues within organisms differ in their response to temperature, but in general physiological activity stops at biological zero, accelerates as temperatures increase above that point, and reach maximum activity somewhere between 77°F and 90°F (25°C and 32°C). At higher temperatures growth efficiency decreases, and at some point the organism is killed. Most microorganisms and cold-blooded animals have a similar response to temperature, although some microorganisms grow at temperatures below 32°F (0°C) and some at temperatures above 194°F (90°C).

DAILY TEMPERATURE DYNAMICS IN THE SOIL

All soils have temperature means and dynamics that are dictated by the climate and weather at that location but modified by the vegetative or mulch cover on the soil surface and the amount of water in the soil. Heating and cooling of soil takes place by the exchange of energy between the soil and the air at the soil surface. Unless covered by snow or a thick cover of mulch, such as a layer of dead leaves or other plant residue, the surface of the soil heats during the daylight hours and cools during nighttime. The amount of heating that takes place each day depends on the amount of direct sunlight that strikes the soil surface. Cooling at night depends on atmospheric conditions, primarily cloud cover that controls the amount of long wave radiation that can pass from the soil to outer space. Cooling is greater on cloudless nights.

In the absence of a sudden change in the weather, such as the passage of a cold front or a rainstorm, the pattern of daily temperature dynamics in the soil is nearly the same for all soils (Figure 3–1). The surface of the soil is warmest by about 3 P.M. and coldest just as the sun rises in the morning. On cloudless days the temperature at and near the soil surface rises rapidly by 10 A.M., but the temperature at a few inches below the surface may be slightly lower than at sunrise. After sunset, about 10 P.M., the temperature at and near the surface has fallen while a few inches below the surface temperatures may be slightly above their 3 P.M. values. Cooling of the soil surface after sunset affects the movement of water vapor in the soil. When the soil surface cools after sunset, warmer air from the sub-soil and air contacts the cooler soil and plant surfaces near the soil surface where the water vapor condenses as liquid we recognize as dew on the grass or soil surface.

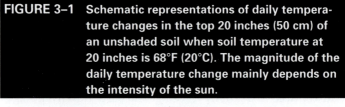

FIGURE 3–1 Schematic representations of daily temperature changes in the top 20 inches (50 cm) of an unshaded soil when soil temperature at 20 inches is 68°F (20°C). The magnitude of the daily temperature change mainly depends on the intensity of the sun.

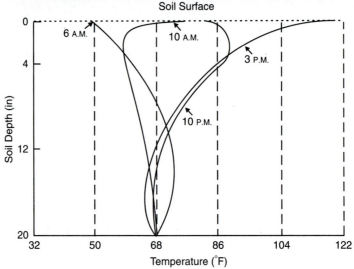

If the temperatures at any depth are averaged for the entire day, all depths have approximately the same average temperature. Note that temperature at a depth of 20 inches (50 cm) does not change during the course of the day and closely approximates the average temperature of all depths above that level. If the weather is becoming warmer from day to day, the temperature at the 20-inch (50-cm) depth will slowly increase each day. A 20-inch (50-cm) depth is a good depth to measure soil temperature to represent the *average* soil temperature for that day because measurements can be taken any time during the day.

The daily pattern of soil temperature dynamics shown in Figure 3–1 is representative of all soils exposed to direct sunlight on days with moderate temperatures, such as the 68°F (20°C) represented in the figure. The magnitude of the temperature dynamics is both much less and much greater than illustrated under some conditions. On cloudy days there is less soil temperature variation because the clouds intercept and reflect direct radiation from the sun away from the earth. A dense cloud cover for the entire day may result in little or no temperature change in the soil.

Radiation from the sun must strike an absorbing surface to be converted to the sensible heat we feel and measure with our thermometers. When radiation directly strikes the soil surface, the sensible heat formed is both dissipated to the air and conducted into the soil. The heated air rapidly rises and in so doing creates vertical currents of air. Light passes through air at different speeds depending on the temperature of the air. As small columns of hot air rise from the soil surface and columns of colder air sink toward the soil surface, the different temperatures bend light waves, and images in the distance appear to be moving or inverted. This optical phenomenon is commonly known as "mirage" and often seen when looking over level land on hot sunny days.

The temperature of the air above the soil surface is often very close to that of the soil surface, but because air is free to move and mix with air above the surface the most extreme temperature variations are found very close to the soil surface. During cloudless days a thermometer placed a few feet above the soil surface may measure air temperature changes of 20°F to 30°F (11°C to 16.7°C) from day to night. The temperature changes at a bare soil surface are usually greater, especially when there is no wind.

The amount of soil temperature change that results from a given amount of radiation depends on several factors. Light-colored soils reflect more of the sun's radiation than dark-colored soils. Reflected radiation (albedo) does not heat the soil or air. Snow has an exceptionally high albedo. Dark-colored soils absorb and convert more of the radiation to sensible heat. Vegetable growers in temperate latitudes can speed the warming of their soil in the spring by spreading black charcoal dust on the soil surface or covering the soil with black plastic to decrease albedo and increase soil temperatures. The amount of soil temperature change that results from the sensible heat generated at the soil surface depends in large part on the water content of the soil. It takes approximately 3.5 times more energy to effect the same temperature change of a soil saturated with water than in a dry soil. The daily amplitude of soil temperature change is much greater when the soil is dry than when the same soil has greater water content. Under equal weather conditions the magnitude of daily soil temperature change is less in soils with a high content of water than in soils containing less water. This is due to the greater specific heat of water (1 calorie g^{-1} water) while the specific heat of dry soil is approximately 0.2 calories g^{-1} (Box 3-1).

BOX 3-1

Heating the Soil

The amount of water in a soil greatly affects how rapidly the soil will warm when heated by the sun. It requires more heat to warm a wet soil than a dry soil. The amount of heat required to change the temperature of a substance is known as the *specific heat capacity* of that substance.

A *calorie* (cal) is defined as the amount of heat required to increase the temperature of 1 g water 1°C. Thus the specific heat capacity of water is 1 cal g^{-1}. When 1 cal of heat is added to 1 g water, the temperature increases by 1°C. Calories of heat added, divided by the specific heat of a substance, equals the amount of temperature, in °C increase that substance will experience. For water that is:

1 g of water/1 cal g^{-1} = 1 cal or 1°C

Most oven-dry soils have a specific heat capacity of 0.2 cal g^{-1}.

When 1 cal of heat is added to 1 g of dry soil the temperature rises 5°.

(1 g/1 cal)/(0.2 cal/1 g) = 5°C

Assume that the oven-dry bulk density of the soil is 1g cm^{-3} and we add water equal to 10% of that weight water (i.e., 0.1 g of water). The volume does not change because the water only moves into the voids in the soil, replacing air that has negligible specific heat capacity. The specific heat capacity of the same volume of soil is now the original 0.2 cal g^{-1} from the soil particles plus 0.1 cal g^{-1} from the 0.1 g of water, for a total of 0.3 cal g^{-1}. The wet soil weighs 1.1 g. but still has a volume of 1 cm^3.

Thus the specific heat capacity on a weight basis is 0.3 cal g^{-1}/1.1 g = 0.273 cal g^{-1}, but the specific heat capacity on a volume basis is 0.3 cal cm^{-3}. Thus the temperature rise after adding 1 cal of heat to 1 cm^3 soil containing 10% water would be:

1 cal/0.3 cal cm^{-3} = 3.33°C

The specific heat capacity of the same 1 cm^3 soil when saturated with water, that is, all the pore space (approximately 50% of the volume, i.e., 0.5 g water) filled with water, would be 0.5 cal from the water plus 0.2 cal from the mineral particles or a total of 0.7 cal cm^{-3}.

When 1 cal of heat is added to 1 cm^3 of the saturated soil, the temperature rise would be:

1 cal/0.7 cal cm^{-3} = 1.43°C

These few examples demonstrate why wet soils are considered cold soils, especially to farmers in temperate areas when they are attempting to plant in the spring of the year. As discussed in other chapters, most microbial activity and chemical reactions are slower in cold soil than in warm soil. The water content greatly affects how rapidly a soil warms under the heat of the sun each day and greatly affects the maximum daily soil temperature near the soil surface.

When at permanent wilting water content, sandy soils, because they have a greater portion of their pore space as large voids and therefore contain less water than more clayey soils, get both hotter and colder during the course of a sunny day than clayey textured soils. Extreme surface temperatures approaching 150°F to 160°F (66°C to 71°C) have been recorded on the surface of dry soils. The temperature at the soil surface does not go higher than this because at those temperatures the amount of radiant energy leaving the hot soil surface equals the amount of radiant energy of sunlight arriving at the soil surface.

Under trees with full leaf cover, the daily extremes of soil temperature are greatly reduced because the sun's radiation forms sensible heat at the leaf surface in the trees and is dissipated from there into the air, never reaching the soil surface. On a hot sunny day we all seek the coolness of a shade tree, thereby protecting our bodies from radiation that is transformed into sensible heat on striking our bodies.

Young seedlings just emerging through the soil surface may be damaged or killed by even short periods of extremely high temperatures. Severe high temperatures immediately at the soil surface can be avoided by keeping the soil moist or insulating the soil surface with a thin covering of dead plant material, that is, mulch. Mulch also reduces evaporation from the soil and is sometimes used to protect young plants emerging from rather dry soil. Although effective, it is difficult and very labor intensive to mulch large areas to protect young plants from extreme heat at the soil surface. Too much mulch can harbor pests and disease organisms that injure small seedlings when soil conditions are moist.

SEASONAL DYNAMICS OF SOIL TEMPERATURE IN TEMPERATE LATITUDES

Figure 3–2 illustrates mean monthly soil temperatures with depth at Lexington, Kentucky, USA, and Table 3–1 shows the mean monthly air and surface soil temperatures over the same five-year period. Mean annual air temperature at the Lexington,

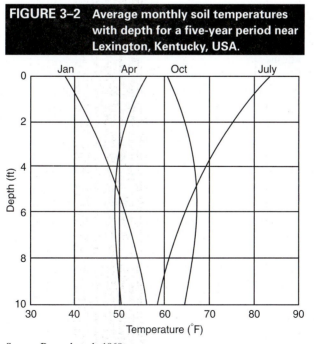

FIGURE 3–2 Average monthly soil temperatures with depth for a five-year period near Lexington, Kentucky, USA.

Source: Penrod et al., 1960.

TABLE 3–1 Mean Monthly Surface Soil and Air Temperatures for 5 years (1952–1956) at Lexington, Kentucky

Month	Soil Surface °F	Air °F	Soil Surface °C	Air °C
January	37.2	35.4	2.9	1.9
February	40.0	39.6	4.4	4.2
March	46.7	44.8	8.2	7.1
April	56.2	55.5	13.4	13.1
May	68.7	64.8	20.4	18.2
June	78.9	74.3	26.1	23.5
July	82.8	78.1	28.2	25.6
August	78.8	76.7	26.0	24.8
September	73.2	70.0	22.9	21.1
October	60.4	57.8	15.8	14.3
November	45.6	44.3	7.6	6.8
December	38.5	37.3	3.6	2.9
Annual Average	58.9	56.6	14.9	13.7

Source: Penrod, Elliot, and Brown, 1960.

Kentucky, site was measured at 56.6°F (13.7°C) while the mean annual soil temperature was 58.9°F (14.9°C) (Table 3–1).

Mean monthly soil temperatures near the surface closely approximate but are slightly warmer than mean monthly air temperatures. Many studies relating mean annual soil temperature to mean annual air temperature have been made, and results differ by only a few degrees. Adding 2°F to 4°F (1.10°C to 1.7°C) to mean annual air temperatures reported in climatic data closely estimates mean annual soil temperature at most locations.

Mean annual soil temperature at any location is nearly the same at all soil depths. At depth the seasonal changes in temperature are less extreme than at the soil surface. At Lexington, Kentucky, the mean monthly temperature near the surface of the soil exceeds 82°F (27.7°C) in July and is less than 38°F (2.8°C) in January. Mean monthly temperature ranged only from 65°F to 50°F (18°C to 10°C) at a depth of 10 feet (3 meters) (Figure 3–2). Also, note the delay in temperature change at the 10-foot (3-meter) depth as compared to the surface temperatures in Figure 3–2. From January to April the mean monthly soil temperatures at the surface increased, but soil temperature below 6 feet (1.8 meters) decreased. A similar lag in seasonal temperature change with depth is present from July to October when the soil temperatures above about 5 feet (1.5 meters) have decreased while temperatures below that depth increased.

Seasonal temperature change with depth can be utilized to help heat and cool buildings in temperate latitudes. By circulating water from the heating and cooling system of a building through heat exchanger pipes buried at a depth of 10 feet (3 meters) or more where the temperature varies only slightly throughout the year, some heating in winter and some cooling in summer is achieved.

ALTITUDE AND SHAPE OF THE LAND INFLUENCE AIR AND SOIL TEMPERATURE

Altitude and shape of the land influence soil and air temperatures and are important to natural vegetative ecosystems and the growing of many crops. Within any hilly or mountainous land area, soil temperatures and air temperatures become cooler as elevation increases. Air temperatures decrease an average of 3.6°F for each 1,000 feet of increase in elevation (6.5°C for every kilometer increase in elevation). Thus snow-capped mountain peaks are present at the equator (Figure 3–3). Also, daily temperature variation increases with elevation. Daily extremes in temperature are greater at high elevations because the air is less dense, and solar radiation is more intense during the day and at night cooling is greater than at lower elevations.

Within temperate latitudes the direction or aspect of the slope affects soil temperature. Slopes that face the equator are warmer than slopes with pole-facing aspects. In Figure 3–4 the pole-facing aspect of the land is vegetated with trees and the equator-facing aspect is vegetated with grass. In temperate latitudes with equator-facing aspects, radiation from the sun is more perpendicular to soil surface than on pole-facing aspects. Thus less of the solar radiation is reflected back to space and more energy is available to heat the soil. This is perhaps most visible during clear days following a snow event in temperate latitudes when snow and ice on the streets on the

FIGURE 3–3   Photo of Mount Kilimanjaro, Tanzania, located within 4° latitude of the equator. The mountaintop, elevation 19,317 feet (5,872 meters) is perennially snow covered; the foreground at approximately 3.280 feet elevation (1,000 meters) has a warm semiarid topical climate.

FIGURE 3–4 Photo of natural vegetation in northern Arizona, USA (35° N. Lat.) showing a hill where the south (equator)-facing slope, to the left on the photo, and level foreground is grass covered, whereas the north (pole)-facing slope is forested.

equator-facing slopes, south-facing slopes in the northern latitudes, melts while snow and ice on the north-facing slopes on the same street may remain for several days.

Although less visible to the casual observer, significant differences related to aspect are also present in the soil during the summer months and significantly affect the growth of plants and other life-forms. Table 3–2 contains data recorded on

TABLE 3–2 Microclimatic and Soil Condition on North and South Facing Slopes in Michigan During the 1957 Growing Season

Measurement	Slope Facing	Apr. 15– June 1	June 2– July 15	July 16– Aug. 23	Aug. 24– Sept. 16	Sept. 17– Oct. 16
Maximum air temp.	North	72.9	79.6	85.8	75.8	70.4
at 20 in (50 cm) (°F)	South	76.7	87.7	95.1	80.3	76.9
Minimum air temp.	North	46.5	58.1	57.1	51.8	39.4
at 20 in (50 cm) (°F)	South	47.7	58.2	57.5	52.2	39.8
Soil temp. at 0.8 in (2 cm)	North	57.6	66.4	70.8	65.6	55.8
(average °F)	South	61.1	69.5	76.1	68.2	60.3
Soil temp. at 7.9 in	North	50.5	61.8	65.5	62.4	55.0
(20 cm) (average °F)	South	55.6	64.9	70.6	65.6	59.6
Soil moisture content at	North	27.4	19.4	13.4	11.6	9.2
0.8 in (2 cm) (average %)	South	19.4	15.4	8.8	10.2	6.8

Note: °C = (°F – 32) × 0.556.

Source: Cooper, 1960.

south- and north-facing (i.e., equator- and pole-facing, respectively) during the 1957 growing season in southern Michigan, USA. Note that both the average soil and near soil air temperatures during the summer growing season are higher on the south-facing slope than on the north-facing slope. Also note that the higher air and soil temperatures have increased evapotranspiration rates and resulted in much lower average soil moisture contents near the soil surface on the south-facing slope than on the north-facing slope.

TEMPERATURE DYNAMICS, ASPECT, AND DAY LENGTH IN THE TROPICS

Daily soil temperature dynamics are the same in tropical latitudes as in temperate latitudes. The only property that is consistently different between soils in temperate latitudes and soils in tropical latitudes is minimal seasonal change in soil temperature within tropical latitudes. Within most of the tropical geographic latitudes, the difference between the mean monthly soil temperature of June, July, and August differs less than 10.8°F (6°C) from the mean temperature of December, January, and February. In most temperate latitudes the mean soil temperature during June, July, and August differs by more than 10.8°F (6°C) from the mean temperature of December, January, and February. Soil scientists use this lack of mean monthly temperature change during the year to define soils located within the tropics by referring to them as having *iso* (i.e., uniformity) soil temperature regimes. Except for a few coastal areas, like in the coastal areas of northern California where the air temperature is modified by ocean currents, these limits correspond quite well with the tropics as geographically defined by the latitudes of Cancer (23.5° N. Lat.) and Capricorn (23.5° S. Lat.) on the globe.

This distinction between tropical and temperate soil temperature dynamics through the year is critical to understanding many of the practices utilized by farmers in growing food crops. In temperate latitudes, seasonal temperatures dictate the growing season for food crops, whereas in tropical latitudes seasonal temperature changes are so small that seasonal temperatures seldom dictate the planting and harvesting of food crops. In the tropics, farmers time their crop growing to seasonal patterns of rainfall rather than seasonal patterns of temperature. Tropical (iso) soil temperature characteristics simply indicate that little or no attention needs to be given seasonal temperature when planting crops.

In tropical latitudes, the sun is nearly perpendicular to level land surface throughout the year, and north- and south-facing slope aspects create little temperature difference. There is a slight tendency for east-facing aspects to be slightly cooler than west-facing aspects. This is because the morning sun first evaporates dew before it heats the soil, whereas after noon, when radiation from the sun is more perpendicular to west-facing aspects, the plant and soil surfaces are moisture free and heat more readily.

Another feature of land in tropical latitudes is day length. Throughout the year each 24-hour day has approximately 12 hours of daylight and 12 hours of darkness, but in temperate latitudes daylight is present for several more hours each day during the summer and reaches 24 hours each day for a short period of time each summer at polar latitudes. Photosynthesis is directly correlated with the amount of sunlight (radiation) a plant receives. The greater duration of sunlight each day during the summer is considered an advantage for rapidly growing crop plants in the temperate latitudes. The uniform day length in the tropics is a consideration for many plant species and cultivars in which seasonal differences in day length (i.e., photoperiod) control flowering and other physiological stages of growth. Some plant species and genetically selected cultivars adapted to long days and short nights in temperate latitudes are not well adapted to the uniform daily light regimes of tropical latitudes.

SOIL TEMPERATURE AND CROP SELECTION

Seasonal changes in soil temperature closely determine the type of food crop that can be grown and when during the year that crop can be grown. Within most tropical latitudes the mean average soil temperature of the three coldest months is less than 10.8°F (6°C) lower than the average temperature of the three warmest months, and these areas are outlined as iso (tropical) in Figure 3–5. Soil temperature has little or no influence on when during the year a crop can be planted in iso areas. There are mean annual temperature differences within the iso regions that determine the type of crops that can be grown. At the higher elevations and thus cooler mean annual soil temperatures in tropical latitudes, cold-adapted crops such as potatoes are often a staple of indigenous people rather than cassava or rice that are more adapted to warmer low-elevation temperatures. The growth rate of most adapted crops within tropical latitudes is slowed by cooler temperatures present at higher elevations, and even the most cold-tolerant crops such as potatoes cannot be grown at elevations where mean annual soil temperatures are below 50°F (10°C) (Figure 3–6).

Within temperate latitudes, the minimum mean annual soil temperature is a reasonable guide to where several important food crops can be successfully grown. Most citrus crops are limited to areas where mean annual soil temperature is above 72°F (22°C) (Hyperthermic in Figure 3–5). Cotton production is limited to areas where the mean annual soil temperature is above 59°F (15°C) (Thermic and Hyperthermic in Figure 3–5). Most corn for grain is grown only where the mean annual soil temperatures are above 47°F (8°C), the cold limit of Mesic temperatures outlined in Figure 3–5. However, corn for silage (i.e., animal food) wherein the entire corn plant is harvested and stored for animal feed can be grown in colder areas. If during 90 consecutive days during the summer months, soil temperatures average above 59°F (15°C), (i.e., Frigid areas in Figure 3–5) corn harvested and stored as silage for animals, wheat, oats, and other cold hardy grains can be successively grown. In the Cryic areas outlined in Figure 3–5 there are less than 90 days each year when soil temperatures are above 59°F (15°C), but some short-season vegetables and extremely cold hardy grains can be grown. The growing of any food crops is seldom possible in areas with permanently frozen subsoil (Gelisols; see the appendix) and designated as Permafrost in Figure 3–5, except for a few vegetables in small protected sites.

High mean annual soil temperatures do not severely hamper the growing of most food crops in most temperate latitudes because an earlier planting date in the spring can be selected for crops that prefer cooler growing seasons.

Soil Temperature Regimes

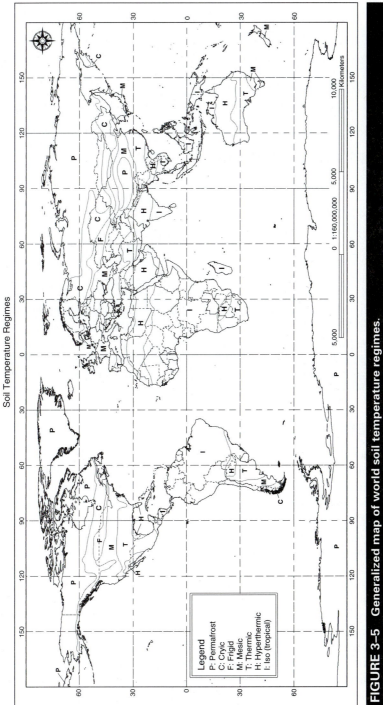

FIGURE 3–5 Generalized map of world soil temperature regimes.

Legend
P: Permafrost
C: Cryic
F: Frigid
M: Mesic
T: Thermic
H: Hyperthermic
I: Iso (tropical)

1:160,000,000

FIGURE 3–6 Photo at high elevations near the equator in Ecuador, SA. Note that the potato fields on the facing slope all end at the same elevation. At that elevation the mean annual soil temperature is 50°F (10°C).

AIR TEMPERATURE OVER THE LAND

During cloudless nights, heat is lost from the soil surface by radiation into space, and the air near the soil surface rapidly cools. Air temperatures immediately above the soil surface become much colder than air only a few feet above the soil surface. Note in Table 3–2 that minimum air temperatures at 20 inches (50 cm) above the soil are only slightly influenced by slope direction. The accumulation of cold air within a few inches of the soil surface is most severe when there is no wind to mix the air, and it may become several degrees colder than temperatures recorded in standard weather stations that are usually placed about 5 feet (1.5 meters) above the ground.

The effect of cold air accumulating within a few inches of the soil surface is particularly detrimental to low-growing crops such as strawberries during the time they are blossoming and setting fruit. If the temperatures around a flowering strawberry plant dip much below 32°F (0°C) for even a few minutes, the flower is killed and no fruit is formed. Strawberry growers often spray water on their plants when such conditions are expected. The water freezes and covers the plant with ice, but the temperature within the ice remains at 32°F (0°C) and the flower is not damaged.

Cold air is heavier than warm air. During the night, cold air near the soil surface flows down hill slopes and accumulates in valleys. This process, known as *cold air ponding* often creates freezing conditions in valleys while adjacent hillsides are frost free. In hilly or mountainous terrain, fruit trees and other crops that may flower and set fruit during seasons that may experience freezing weather conditions are most often planted on sloping land well above the valley bottom to reduce the danger of freezing.

FROST-FREE SEASON

Although mean annual soil temperature is a regional guide to land suitable for many of our major food crops, many plants are susceptible to only a few minutes of freezing conditions above the soil surface. Not all plants have the same tolerance to freezing. Although 32°F (0°C) is the freezing point of pure water, many plant tissues are able to withstand somewhat lower temperatures without cell damage. Plants are most susceptible to frost damage when they first begin growth or when they are in the flowering stage of growth. In temperate latitudes, the planting date of most crops is governed by the probability of freezing conditions occurring at that site. The number of frost-free days that can be expected each year is known as the frost-free growing season. A minimum of 100 consecutive days without freezing conditions is necessary for most food crops, although some crops can be grown in less time and many require longer frost-free seasons. Considerable scientific effort has been devoted to developing cultivars of major food crops that can complete their growth in as short a period of time as possible so as to allow production in colder climates by reducing the risk of frost damage.

In the temperate latitudes, it can be estimated that the probability of a killing frost is minimal when the mean monthly temperature is above 55°F (12.8°C). However, each year is a gamble because the exact date of a killing frost, either in spring or fall, cannot be accurately predicted.

INFLUENCE OF TEMPERATURE ON WATER USE IN FOOD PRODUCTION

One of the most significant relationships associated with temperature is its influence on water use by plants. Chapter 2 (Table 2–3 and associated discussion) introduced the relationship of temperature and potential water consumption (potential evapotranspiration) by plants. The interaction of average seasonal rainfall distribution and average seasonal temperatures has a profound effect on when farmers plant food crops throughout the world.

An overview of average seasonal temperature and precipitation patterns in various locations illustrates how these physical parameters interact to influence human food production practices and provides insight into the social and economic patterns in different parts of the world. Be aware that patterns of temperature and precipitation are greatly altered by topography, and therefore local land management practices often differ greatly within each of the regional conditions discussed.

Measurements of temperature and rainfall have been recorded for several years in many places throughout the world. In parts of the world only sparse records are available. Hargreaves and Samani (1986) have assembled one of the most complete collections of climatic data that includes monthly averages of temperature, precipitation (ppt), and calculations of potential evapotranspiration (PET). Average monthly temperatures are most often determined by adding the daily maximum temperature to the daily minimum temperature and dividing by 2 to determine the daily average temperature. The average daily temperatures during the month are then added and

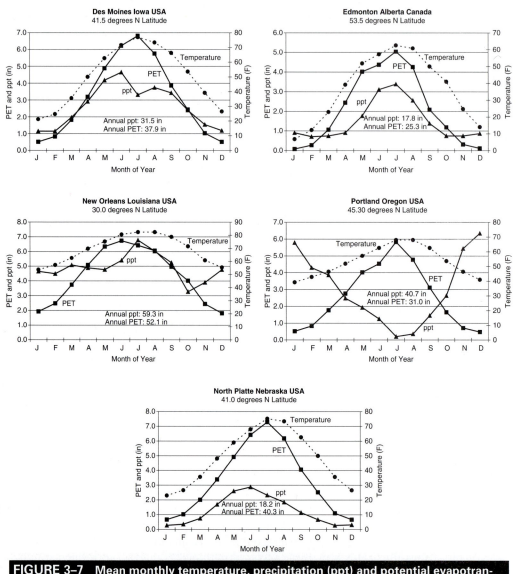

FIGURE 3–7 Mean monthly temperature, precipitation (ppt) and potential evapotranspiration (PET) diagrams for ten selected locations.

divided by the number of days in the month to determine the mean monthly temperature. Figure 3–7 presents graphs of mean monthly temperature, ppt, and PET constructed from data published by Hargreaves and Samani (1986) for selected locations in various parts of the world. Annual average ppt and PET values are also given for each location. When viewing these graphs, remember that PET values represent actual evapotranspiration only when water is available to the plants and there is a complete vegetative cover of actively growing plants. During the time that vegetation is absent or dormant due to freezing conditions or cultivation, there is little or no *actual evapotranspiration.*

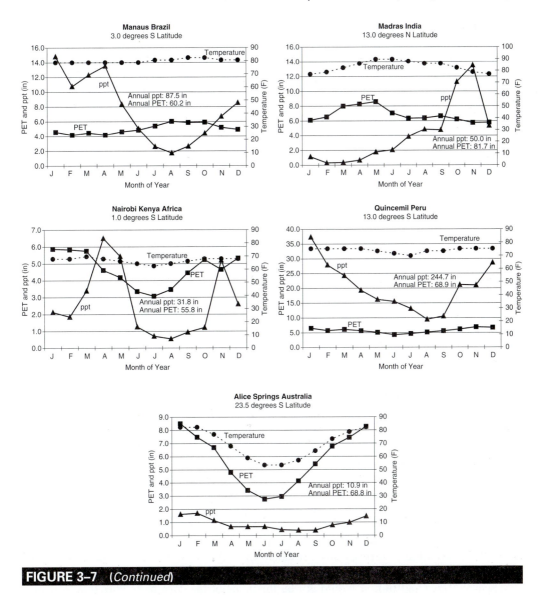

FIGURE 3–7 (*Continued*)

Land where in normal or average years there is adequate moisture for growing food crops throughout the season of the year when temperatures are suitable for such crops is considered *humid*. Des Moines, Iowa, in the heart of the corn-growing area in the U.S. Midwest has a climatic pattern representative of what temperate latitude cultures consider the best for corn, one of the most utilized human food crops. At Des Moines, average monthly air temperatures reach about 55°F (12.8°C) by about May 1. Corn is planted in late April or early May when it is a reasonably safe bet that the danger of a killing frost is minimal by the time the corn seedlings emerge, although each year that is always a gamble. Average monthly temperatures remain above 55°F (12.8°C) until mid-October, providing approximately 160 frost-free days, but individual years vary from as little as 140 to as much as 200 days. At Des Moines, average rainfall is less than the optimum demand, that is, PET, for water throughout the entire period the corn is growing. Long daylight hours and a mean July temperature approaching 80°F (26.7°C) indicate that a rapidly growing crop like corn will need more than 6 inches (15.2 cm) of water during the month. On average only a little more than 3 inches (7.6 cm) of precipitation is expected in July. Several factors have to be considered to account for the success of growing corn in Iowa with the apparent lack of water needed for corn production. Actual water demand is equal to PET only when there is a complete canopy of green leaves. Thus at Des Moines there is little or no loss of water by evapotranspiration from about November through March due to freezing conditions. Actual evapotranspiration loss of water will not equal PET in a corn crop until about a month after it is planted because the small corn plants do not form a complete canopy. Actual evapotranspiration demand also becomes less than PET as the corn ripens. Even after considering reduced moisture requirements in the early stages of growth and a period in the fall as the corn ripens, it is clear that during average years there is not enough rain during the growing season to meet the PET demands of the growing crop. During the summer months, approximately 4 to 5 inches (10.2 to 12.7 cm) of water must come from plant-available water stored in the soil. This amount of water can be extracted from a depth of approximately 20 to 25 inches (50.8 to 63.5 cm) in soils of silt loam texture that retain approximately 0.2 inches of plant-available water per inch (0.2 cm per cm) of soil depth (Table 2–2). Deep silt loam soils are prevalent near Des Moines and throughout the major corn-growing areas of the Midwest of the United States, and corn roots easily access 25 inches (63.5 cm) of soil depth in most soils.

Also note that at Des Moines the annual ppt is less than the annual PET. Thus on an annual average basis it would seem there would not be enough water to produce excellent crop growth. The key explanation of satisfactory corn growth at Des Moines lies in the fact that water is reliably stored in the soil during the fall, winter, and early spring when almost no plants are actively growing in a cultivated cornfield. This is true for almost all of the major corn-growing areas of the Midwest.

Also, note that there is an average decrease in rainfall after June in Des Moines. The relative dryness and cooler air temperatures late in the growing season have an added advantage by lowering the moisture content of the corn grain before harvest. This is an advantage that reduces the cost of drying corn and other grains such as soybeans to moisture contents necessary to prevent spoilage of the grain while in storage.

If cold temperatures did not cause the vegetation to die so the precipitation received during the winter could enter the soil and be stored for the next growing season, Iowa and Illinois would not be the heart of the corn belt. Note that the native vegetation of the area before European settlement was grassland grazed by herds of buffalo, elk, antelope, and deer and frequently burned by wildfires as the grass dried in the fall. Corn is a grass species, and its growth pattern and requirements are much the same as those of native grasses and deciduous trees in the area.

At the cold fringe of food crop growing in northern temperate latitudes, we can look at climatic conditions at Edmonton, Canada. At Edmonton we find that average monthly temperatures exceed 55°F (12.8°C) only one month each year. Clearly there is not enough time to grow corn for grain. There are approximately 3 months (90 days) when average temperatures exceed about 48°F (8.9°C), and although freezing conditions of 32°F (0°C) may occur near the beginning and end of this period, cold-tolerant oat, barley, wheat, and canola crops are possible. Like Des Moines there is inadequate rainfall to meet the PET requirements during the crop-growing season, and the crops must rely on moisture stored in the soil during the winter months.

Moving to near the warm extreme of temperate latitudes on land with generous rainfall, we can view the climatic conditions in New Orleans, Louisiana, USA. There we find that mean monthly temperatures are warm enough that freezing conditions can be expected during only about two months in average years. This is enough frost danger that perennial citrus trees are endangered, and therefore citrus production is limited to still warmer conditions found in southern Florida and Texas and lower elevations in southern Arizona and California within the United States. Corn, wheat, and other grain crops are grown in Louisiana. However, the relatively high amount of rainfall that can be expected each month is a disadvantage because of disease and lack of a dry period during which the grain can mature and attain low-moisture contents suitable for storage. The long growing season is suited to growing cotton. Although much cotton is grown, moist conditions often interfere with cotton harvest as expressed in the song lyric as "When those cotton balls get rotten, you can't pick very much cotton." The relatively high temperatures and adequate rainfall through much of the summer, coupled with the relative level land suitable for retaining water without much runoff, are ideal conditions for growing rice, often planted in rotation with soybeans whose seeds can tolerate moist conditions.

Land where in average years there is not adequate moisture for growing food crops during a portion of the time that temperatures are suitable are considered *semiarid.* In temperate latitudes, this is where a combination of rainfall plus available soil water fails to provide enough water for 90 or more consecutive days during a season when temperatures are adequate for crop growth.

Two rather distinct semiarid conditions can be recognized in temperate latitudes. The most prevalent semiarid condition is illustrated by conditions at North Platte, Nebraska, USA. Although PET exceeds ppt each month of the year, there is essentially no actual evapotranspiration during the winter and precipitation is in the form of snow. As temperatures warm in the spring months, the snow melts and enters the soil as water that is available for plants that then begin to grow. However, the amount of winter precipitation is very small, and in average years there is not enough available water stored in the soil to adequately meet the PET needs of crops during the warm summer months without supplemental irrigation. Crop production is possible without irrigation in some years with above-average rainfall, but as the early European settlers found to their great hardship in the 1930s, drought conditions frequently prevail.

One practice that farmers adopted in areas like North Platte, and still practice in limited areas where irrigation is not available, is known as *dry-land fallow* or *summer fallow.* Farmers practicing dry-land fallow farming plan to grow only one crop every two years. During the summer when no crop is to be grown, the surface of the soil is cultivated to prevent the growth of weeds that would transpire water. This allows much of that summer's rainfall to be stored as available water in the soil. The extra stored available water usually makes it possible to grow a crop the next year. The practice of cultivating the soil to control weed growth is sometimes called a "dust fallow," so named because of the wind erosion that often occurs on dry soil not protected by vegetation. Extensive areas of dust fallow contributed to the intensity of dust storms in the Dust Bowl during the 1930s when for several years rainfall was below normal and clouds of dust from fields in western Nebraska, Kansas, Oklahoma, and other areas with climatic conditions like those at North Platte darkened the skies all the way to the East Coast of the United States. Many farms in these areas were abandoned during that period and allowed to return to native grassland.

A rather contrasting scenario of seasonal rainfall and temperature is present on some lands in temperate latitudes where there is an abundance of rainfall in the winter and extreme drought in the summer. Portland, Oregon, USA, is an example of this condition, often referred to as a "winter rain" area. Winter temperatures in Portland are mild compared to North Platte, and many cold-tolerant plants can be planted each year, but they must mature prior to the extreme drought conditions that occur during the summer or receive supplemental water from irrigation.

Within the tropics, the uniformity of temperature throughout the year presents a temperature/moisture scenario that contrasts to conditions in temperate latitudes in that there is no seasonal cold period during which there is little or no evapotranspiration. Seasonal patterns of rainfall determine when crops can be grown.

A few areas in the humid tropics receive more precipitation each month than is required to meet potential evaporation demands. These lands can be referred to as *perhumid.* One of these areas on the eastern foot slopes of the Andean Mountains in Peru is Quincemil. Although a frost-free condition with excess rainfall each month may seem like an ideal location for growing food crops, it is quite the contrary. In the constantly humid conditions, disease and insects ravage most food crops. It is nearly impossible to dry grain crops, and humans rely largely on fruits and other indigenous plants for food. The perpetual wetness makes it difficult to burn jungle vegetation for slash-and-burn farming techniques. Roads and other infrastructure are difficult to build and maintain. Most perhumid areas are sparsely populated.

Much of the land in the tropics is somewhat less humid and has one to three months each year during which PET exceeds average precipitation. Such areas are commonly known as the *humid* tropics. It is possible to grow crops almost anytime during the year with two and in some places three consecutive crops grown each calendar year. Grain crops are usually not planted at the beginning of the relatively dry period of the year. The relatively dry period permits the harvest of grain crops that can adequately mature in the relative dryness but is not severe enough to halt the growth of perennial crops. It is also a time when jungle vegetation can be burned for slash-and-burn management.

Semiarid lands in the tropics are also known as *seasonal rainy* lands. Semiarid land in the tropics receives adequate moisture for growing one or two crops each year, but adequate moisture is not available to grow crops during more than 90 consecutive days each year. Semiarid lands in the tropics are somewhat unlike semiarid land in temperate latitudes. In the tropics, growing seasons are not limited by the threat of freezing temperatures, and PET rates are high every month of the year. Although the PET rates remain nearly constant throughout the year, they are somewhat lower than for equivalent monthly temperatures during summer months in temperate latitudes because of shorter day lengths in tropical latitudes.

Several rather contrasting seasonal precipitation patterns interact with the nearly constant PET rates to determine crop-growing seasons in the semiarid tropics. Unlike conditions in temperate latitudes where at the time of planting food crops the soil is fully supplied with water from precipitation during the cold winter months when the vegetation is not transpiring, soils in the semiarid tropics are almost entirely devoid of stored water at the time of planting. Farmers, in anticipation of a period of adequate rainfall, most often plant at or slightly before the time of the first rain events of the expected rainy season. If the first rain events of the anticipated rainy season cause germination but continued rainfall is inadequate for only a few days, the shallowly rooted seedlings often die and expensive replanting is required. This is a common risk experienced by farmers in semiarid tropics seldom experienced in temperate latitudes.

Manaus, Brazil, and Madras, India, are examples of semiarid (seasonal rainy) land in the tropics. Although similar in many respects, they differ in several ways. At Manaus the rainy season begins in November and continues through June. This provides approximately eight months of more than adequate moisture to provide for the transpiration needs of growing crops. It is possible to grow two food crops during this time, although the rainy conditions often hamper grain crops that mature during the rainy period and are best planted during the rainy period and harvested at the end of the rainy period. In Madras adequate rain is present for only about two months each year. Rain in excess of PET during this period is stored in the soil and used by the crop at the end of the growing season, thereby providing a minimum of about three months to grow food grains.

Nairobi, Kenya, is another example of some semiarid lands in the tropics where two seasonal rainy periods are expected to occur each year. The major rainy period near Nairobi is April and May. Planting is done in dry soil in anticipation of adequate rain during the rainy months with harvest in approximately 100 days. Like Madras, reliable rainfall is critical following planting. In Nairobi there is a secondary, or minor, rainy season in November. This is a much shorter rainy period than during the primary rainy season. Again there is the need for the rains to continue in a timely fashion after planting and be enough in excess of PET so enough water is stored in the soil to mature the crop. Corn is a preferred crop in the major rainy period. Sorghum and millet are more capable of producing satisfactory grain yields even when stressed by lack of moisture and are commonly grown in the secondary rainy period or during the major rainy period in years where the onset of the rainy season is late. If it is a good year, a crop is successful in the secondary rainy season, but there is more of risk of crop failure than during the primary rainy season.

In all tropical land with a seasonal rainfall pattern, food crops have a severe risk of crop failure because unlike conditions in temperate latitudes where water is stored in the soil during the winter, there is little or no available water in the soil at planting time. If timely rainfalls do not follow the time of planting, the seedlings often die from lack of water because of high temperatures and PET rates. Each year rainfall patterns and amounts deviate from climatic averages. We should all remember that an average means that half of the time conditions are greater than the average value and half of the time conditions are less than the average. Each growing season farmers risk their investment in time and money when planting a crop.

All land where the soil is nearly devoid of available water more than six months each average year and a combination of precipitation and available water in the soil fails to provide water for 90 consecutive days during the time temperatures are adequate for crop growth is considered *arid*. In practical terms, arid land is where irrigation is necessary to produce most human food crops. An extreme example of arid land is Alice Springs, Australia. Arid land is present both in temperate and tropical latitudes, and the Alice Springs data were selected to remind you that the calendar months of summer and winter in southern temperate latitudes is the reverse of northern temperate latitudes. With unrestricted radiation due to a lack of clouds, most arid lands have a high potential for food production and are relatively free of disease and pestilence that bedevil food crops in more humid areas, but they must have a reliable source of irrigation water. Irrigation for food crop production in arid regions is discussed in later chapters.

GEOGRAPHIC LOCATIONS OF SEASONAL MOISTURE FOR CROPS

Figure 3–8 outlines the locations in the world where seasonal distribution of temperature and moisture dictate when crops can be grown. Be aware that the small scale of Figure 3–8 permits only a very generalized representation of seasonal moisture and temperature locations. Within all of the areas outlined there are significant areas that have contrasting conditions. This is especially true in mountainous areas where local conditions of elevation drastically affect both temperature and rainfall patterns.

Humans throughout the world have adopted similar food production practices to cope with seasonal patterns of rainfall and temperature in their quest for sustained food production. In Figure 3–8, areas where cold temperatures preclude almost all food production are designated TC (Too Cold) and inhabitants rely on migrating herds of animals, fish, or imported food for survival.

Extensive arid areas are located in the western half of the United States, much of Africa, central Asia, central Australia, and southern South America. In arid areas, crop production is confined to land where irrigation is possible and humans rely on cattle, sheep, and goats that can graze on sparse vegetation and provide meat for human consumption.

In semiarid areas, various combinations of annually grown food crops during the rainy season, often augmented by irrigation where possible, and cattle grazing are the most common farming practices. Semiarid areas are prevalent adjacent to more arid areas in the western United States, central and Eastern Europe, Canada, and Mexico. The tropics of central Africa, central South America, northern Australia, and portions of southeastern Asia, most notably India, are semiarid. In semiarid areas of temperate latitudes, it is necessary to harvest hay during the summer to feed the cattle during some winter months. In the tropics where the vegetation is not covered by snow, grazing animals often rely on dead grass and often experience weight loss during the dry season. Severe rates of mortality often result in years with unusually severe dry seasons.

A rather unique seasonal timing of winter rain and moderate winter temperatures exists in some temperate latitudes. Most of the late summer months are extremely dry, but the winter rains are reliable and recharge the soil with enough water for some annual crop production. This pattern of winter rainfall and summer drought is in a belt from Southern California through central Oregon and Washington and also in eastern Washington, Oregon, and parts of Idaho in the United States. Similar areas are also present in many areas in southern Europe and the Near East and often referred to as a Mediterranean climate. A small area of winter rain in Chile is prized for grape and wine production, as are southern France, Italy, and Spain and in Northern California. In winter rain areas, some grain crops that can grow slowly during the moderately cold winter are planted in the fall and mature in the spring and early summer months while using available moisture stored in the soil. A wide variety of fruits and vegetables are produced in these areas. Grazing of meat animals that can utilize native grasses is often practiced in these areas.

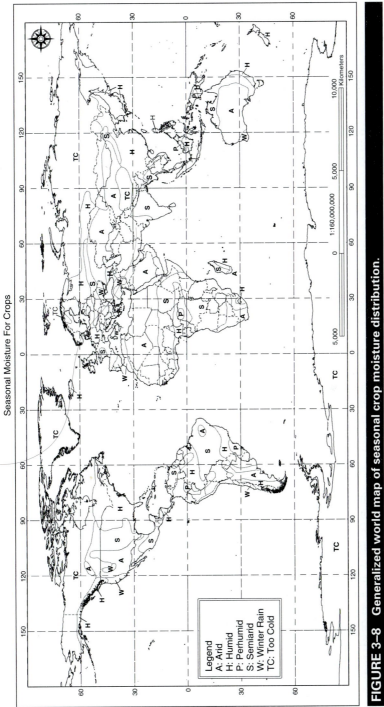

FIGURE 3–8 Generalized world map of seasonal crop moisture distribution.

Source: Van Wambeke, 1981, 1982, 1985.

Humid conditions in northern temperate latitudes that permit growing annual crops of corn, wheat, rice, and other cereal grain crops without irrigation during the summer are present in much of Europe, the eastern United States, and temperate latitudes in eastern Asia. Similar climatic conditions and food production systems are also present in parts of South America, southern Africa, New Zealand, and Australia in southern temperate latitudes. Although grain production is the primary use of these lands, the raising of both milk and beef cattle, hogs, and poultry is extensive in these areas. Intensive feedlot operations to produce meat animals are often located in these areas to take advantage of the locally grown grain.

In the topics, humid conditions are present in much of Indonesia, southern India, central Africa, and central South America. Many grain crops are grown in these humid areas, but rice production has tended to be most extensive, especially on the wetland areas. The other grain crops are more frequently grown in the semiarid tropical areas. In humid areas, cattle are commonly grown on confined and managed pastures rather than freely grazing on native vegetation. In temperate latitudes, it is necessary to harvest vegetation as hay or silage during the summer to feed the animals during the winter when no grazing is available. In humid areas of the tropics, supplemental forage is seldom necessary. In most years irrigation is not required, but many farmers in humid areas have some irrigation capability to protect their crops during periods of drought.

Interspersed within the humid areas are extensive areas of wetland where cereal crop production is only possible when excess water is removed and the water table lowered by engineered drainage systems. Contiguous areas of wetland are too small to indicate in Figure 3–8, but engineered drainage has been installed and maintained as a rather common practice to facilitate the growing of crops on wetlands throughout most humid areas.

Perhumid areas in the tropics are densely forested and have sparse human populations. Extraction of valuable timber and some fruit production does take place, but in general subsistence farming is the primary way of life.

PERSPECTIVE

Temperature is one of the most difficult and expensive properties of land and soil to alter or control. Humans have concentrated their food production on land that is located where temperatures are compatible with food crop production. It is impossible to be specific as to temperatures necessary for the growth of all the possible food plants. Human selection and other modern plant breeding technologies have further complicated such a task by genetically altering most common food crops. Various plant breeding and genetic engineering technologies have created cultivars of all major food crops that are better adapted to specific environmental conditions than the original species. Cultivars and hybrids adapted to shorter growing seasons, longer growing seasons, cool night temperatures, specific disease tolerance, and specific insect tolerances are continually being created. Given adequate water, food crops are grown in the warmest of land, but approximately 100 consecutive frost-free days can be considered as a reasonable minimum growing season requirement for most major food crops. Some vegetable crops can be grown in areas with less frost-free periods, especially in the long daylight hours of polar latitude summers.

A global view of the farming practices humans have adopted to sustain food production must consider the seasonal interaction of temperature and reliable moisture. Although water supply can be manipulated by irrigation of arid and semiarid land and engineered drainage of wet lands, humans can alter temperature only at great expense. Remember also that any discussion of temperature/moisture conditions relies on average conditions that seldom occur in any individual year. Also, localized conditions of temperature and rainfall patterns that contrast with regional averages occur in all areas of the world and cannot be identified on generalized maps.

LITERATURE CITED

Cooper, A. W. 1960. An Example of the Role of Microclimate in Soil Genesis. *Soil Science* 90:109–120.

Hargreaves, G. H., and Z. A. Samani. 1986. World Water for Agriculture. International Irrigation Center, Department of Agricultural and Irrigation Engineering, Utah State University, Logan, Utah, p. 617.

Penrod, E. B., J. B. Elliot, and W. K. Brown. 1960. Soil Temperature Variation (1952–1956) at Lexington, Kentucky. *Soil Science* 90:275–283.

Van Wambeke, A. 1981. *Soil Moisture and Temperature Regimes — South America.* Cornell University and USDA-SCS Soil Management Support Services Tech. Monograph #2. Ithaca, NY.

Van Wambeke, A. 1982. *Soil Moisture and Temperature Regimes — Africa.* Cornell University and USDA-SCS Soil Management Support Services Tech. Monograph #3. Ithaca, NY.

Van Wambeke, A. 1985. Soil Moisture and Temperature Regimes-Asia. Cornell University and USDA-SCS Soil Management Support Services Tech. Monograph #9. Ithaca, NY.

CHAPTER REVIEW PROBLEMS

1. What is biological zero?
2. At what time during a clear day is the soil surface temperature the hottest? The coldest?
3. Compare how day length and seasonal temperature changes differ between the tropical and temperate latitudes of the earth.
4. Where on the landscape does cold air accumulate at night? Why?

CHAPTER

4

Elements Essential for Life

Only a few of the elements present in the air and minerals of the earth are known to be essential for living organisms. These elements exist as inorganic compounds in water, air, and soil. They pass through a complex of organic compounds as life-forms grow, reproduce, die, and decompose, to again become inorganic compounds in the air or minerals in the soil or the waters of the planet. Finite quantities of each life-essential element are present on planet Earth. Although elements combine to form numerous compounds, the absolute total quantity of each element is limited. Although some of the essential elements are present both in air and soil minerals, plants are able only to acquire some essential elements from the air and others from the soil. Water is required for all the essential elements to move through all forms of life.

In this chapter we examine where the essential elements naturally originate in air and soil. In Chapter 5 we examine how they move through the complex organic and inorganic interactions in soil, land, air, and living organisms.

Table 4–1 lists only 17 of the more than 100 known elements on earth. All of the elements discussed and perhaps a few others are present in plants and animals. Silicon and aluminum as mineral compounds with oxygen are the major elements in soil. They are known to be in plant and animal tissue but are not considered essential. Each of the other 15 elements listed in Table 4–1 is known to be essential for plant and animal tissue. Note the extremely different relative proportions in which these elements are present in soil minerals, air, plant tissue, and the human body. Neither plant nor animal life can create elements but must secure them from the air, water, or soil minerals. The three most abundant elements in plant and human tissue are carbon, oxygen, and hydrogen, all obtained by plants from air and water. The natural source of the other essential elements, except nitrogen and some sulfur, is soil minerals. Humans and other animals obtain the elements they need by consuming plant and animal tissue.

LAW OF THE MINIMUM

The amount of each essential element required by living organism differs between both species of organism and tissues within each organism. An insufficient quantity of any one of the essential elements disrupts the biological functions and rate of organism growth. This universal truth is known as the *law of the minimum*. This law can be

TABLE 4–1 Approximate Average Elemental Composition of Soil, Air, Plant Tissue, and the Human Body.*

Elements (Abbreviation)	Soil Minerals	Air	Plant Tissue	Human Body	Natural Source
	%	%	%	%	
Carbon (C)	0.6	0.008	45	18	Air
Oxygen (O)	46	21	43	65	Air/Water
Hydrogen (H)	2.6	0.01	6	10	Air/Water
Calcium (Ca)	3.6	—	2	2	Soil
Nitrogen (N)	Trace	78	1.5	3	Air via Soil
Potassium (K)	2.7	—	1.4	0.35	Soil
Phosphorus (P)	0.1	—	0.5	1	Soil
Magnesium (Mg)	2	—	0.4	0.05	Soil
Sulfur (S)	0.06	—	0.1	0.25	Soil/Air
Iron (Fe)	0.09	—	Trace	0.04	Soil
Manganese (Mn)	Trace	—	Trace	—	Soil
Zinc (Zn)	Trace	—	Trace	—	Soil
Copper (Cu)	Trace	—	Trace	—	Soil
Boron (B)	Trace	—	Trace	—	Soil
Molybdenum (Mo)	Trace	—	Trace	—	Soil
Silicon (Si)	28	—	Trace	—	Soil
Aluminum (Al)	8	—	Trace	—	Soil

*The natural source from which plants obtain each element is indicated as air, soil, or water.
Trace, less than 0.1%.
—, minute quantities.

extended to include not only essential nutrient elements but also life-essential temperature and moisture, discussed in Chapters 2 and 3.

In Figure 4–1 the *law of the minimum* is depicted as a lake. The water level in the lake represents the growth of plants and is illustrated as being maintained by a dam. Each spillway in the dam is represented as an essential nutrient element, with temperature and moisture requirements included as separate spillways. On the right-hand side of the figure, a 100% level has been established to represent the desired lake level. The desired lake level represents the potential rate of plant growth possible with the genetic capability of the species being grown and the solar radiation available at that location. The present level of the lake is represented at 50% of its potential level because copper is not in adequate supply. If the amount of copper available to the plant was increased, by raising that spillway to the 100% level, insufficient amounts of iron and sulfur would allow the lake level to rise only to the 75% level. Note that elemental supply above the 100% level, the maximum genetic capability of the plant with the solar radiation available at the site, does not raise the level of water in the lake. Increasing the availability of any or all of the essential elements or the optimum requirements for temperature and moisture does not overcome the genetic limitation to growth. In fact, excess amounts of an essential nutrient element can be injurious to plant growth and productivity.

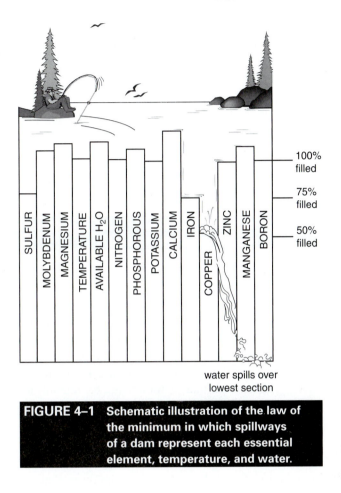

SULFUR
MOLYBDENUM
MAGNESIUM
TEMPERATURE
AVAILABLE H_2O
NITROGEN
PHOSPHOROUS
POTASSIUM
CALCIUM
IRON
COPPER
ZINC
MANGANESE
BORON

100%
filled

75%
filled

50%
filled

water spills over
lowest section

FIGURE 4–1 Schematic illustration of the law of the minimum in which spillways of a dam represent each essential element, temperature, and water.

BALANCE

Figure 4–2 conceptualizes plant response to the availability of essential elements. Relative plant growth and vigor are represented on the vertical axis, and five conditions of plant response are identified. If there is far too little of any one, or more than one, essential nutrient element, there is complete *starvation* and the plant may die. If the availability of any essential nutrient element is too low to fully meet growth and reproduction needs of the plant, a *deficiency* condition exists and plant growth and vigor are less than optimal. When all essential nutrient elements are available in adequate quantities, *optimal* growth is obtained. Too much of a good thing, in this case an overconcentration of some element, may cause poor plant growth. This is termed a *toxic* condition, and the plants affected will show abnormal growth. Finally, if the concentration of some essential element or elements becomes so high that physiological functions of the plant are disrupted for a long period of time, the plant dies (i.e., *lethality*). Toxic and lethal conditions seldom occur because of an overabundance of the major elements but primarily result from high concentrations of elements required only in trace amounts (see Table 4–1) and/or the presence of certain nonessential elements. Toxic and deficiency conditions most frequently occur when the plant is stressed by adverse weather conditions.

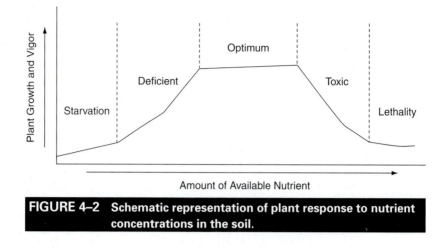

FIGURE 4–2   Schematic representation of plant response to nutrient concentrations in the soil.

Each segment of relative plant growth in Figure 4–2 has a rather broad range. Most plants are able to survive over a wide range of essential nutrient element availability. For example, plant growth may appear optimal, but if the availability of any essential element is near the low end of the optimal stage, the concentration of that element will be less than in plants where the availability of that element is greater. Although the growth of the plant may be little affected, a difference in elemental composition may be great enough that its food value for animals is compromised. Not all healthy plants or fruits of the same species have exactly the same chemical composition.

ESSENTIAL ELEMENTS FROM THE AIR

Strictly on a mass basis, air and water provide the three elements—carbon, hydrogen, and oxygen—that contribute most to the weight of plants and animals.

Carbon, the most abundant element in plant tissue, is present in the air as the gas carbon dioxide (CO_2). Note in Table 4–1 that the percentage of carbon in the air is very small. Carbon is present in some minerals and organic compounds in the soil, but plants do not ingest carbon through their roots systems, except for a few species and then only in minute quantities. Plants obtain carbon as CO_2 gas in the air through openings known as *stomata* in the leaves and other green tissue. Lack of carbon is usually not considered a constraint to plant growth, although it is well known that increasing the CO_2 concentration in the air often enhances growth rate of many plants.

Oxygen, as O_2 gas, constitutes 21% of the air. Oxygen is also a component of CO_2 in the air and water (H_2O). Thus plants obtain oxygen as they ingest oxygen both as CO_2 from the air and water from the soil.

Hydrogen in the air is primarily a component of water vapor reaching the land and plant surfaces as liquid water. Most water is taken into the plant through roots in the soil. Lack of hydrogen as an essential element is best considered as a lack of water in the soil.

Nitrogen is universally abundant in the air over all lands because N_2 gas nitrogen constitutes 78% of the earth's atmosphere. However, plants are not able utilize N_2 from the air. Plants obtain the vast majority of the nitrogen they need through their roots as either nitrate (NO_3^-) or ammonium (NH_4^+) ions in the soil water. Small amounts of nitrogen associated with small organic molecules in the soil may enter the roots of some kinds of plants, and very minor amounts of nitrogen may be absorbed through the plant leaves wetted with water containing nitrogen. Some nitrogen from the air enters the soil dissolved in rainwater, but most enters the soil through specialized *nitrogen-fixing microbes* that live in and on the surface of the soil. Despite the abundance of nitrogen in the air, it is frequently a limiting element in the growth of plants. The transfers of gaseous nitrogen in the air to organic forms of nitrogen in microbes in the soil and then the conversion to inorganic nitrate and ammonium forms available for uptake by plant roots are discussed later. Few minerals in the soil contain nitrogen, and in many soils nitrogen-bearing minerals are totally absent. Although plants obtain almost all nitrogen through their roots in the soil, air is the primary source of nitrogen.

ESSENTIAL ELEMENTS IN THE SOIL

With the exception of carbon, oxygen, hydrogen, nitrogen, and some sulfur that is also present in the air, soil minerals are the primary source of all the other essential elements. Each of the life-essential elements is present in many chemical forms in soil minerals. Most of the chemical forms in soil minerals are unavailable to plants, and thus the total amount of each essential element in the soil is of little or no value in identifying the ability of a soil to provide essential elements of plants. With very minor exceptions, essential elements must be in inorganic ionic chemical forms in soil water and pass into the plant root as the plant absorbs water from the soil. Table 4–2 lists the primary plant-available forms of each essential element that are able to enter a plant root.

It is not possible to identify what percentage of the total amount of each element present in the soil is in a plant-available form. Many conditions in the soil affect the chemical form of each element while in the soil. Different elements are affected differently by conditions in the soil and oscillate from their plant-available ionic form to unavailable organic and inorganic forms in response to soil water content, pH (acidity or alkalinity) of

TABLE 4–2	Plant-Available Forms of Major Nutrient Elements Present in Soils
Element	*Plant-Available Forms*
Nitrogen	NO_3^-; NH_4^+
Phosphorus	$H_2PO_4^-$; HPO_4^{2-}
Potassium	K^+
Calcium	Ca^{2+}
Magnesium	Mg^{2+}
Sulfur	SO_4^{2-}
Iron	Fe^{2+}
Manganese	Mn^{2+}
Zinc	Zn^{2+}
Copper	Cu^{2+}
Boron	$B(OH)_3$; $B(OH)_4^-$
Molybdenum	MoO_4^{2-}

the soil water, temperature, and microbial activity rates. The plant-available portion of the total amount of an element present in the soil almost constantly fluctuates in response to weather conditions. Most studies estimate that less than 10% of the total amount of any essential element in the soil is in a plant-available form. It is probable that less than 1% of the total content of some essential elements in soil are available for plant growth at any one time. Thus 90% to 99% of the total amount of each essential element present in the soil is present as organic compounds or inorganic minerals unavailable to plants.

The plant-available forms of the essential elements are present as inorganic ions in the water that is in contact with the plant roots (Box 4-1). Those elements that are available as cations (identified with $^+$ superscripts in Table 4–2) are also attracted by the negative charges, called the cation exchange capacity (CEC) of the clay and organic matter particles in the soil (Box 4-2). As pictured in Figure 4–3, these cations bounce off the clay or organic matter surfaces into the surrounding water and return much like a basketball being dribbled. Plant roots are able to capture these elements when they are in the soil solution by giving up a positively charged hydrogen ion (H^+) for each charge on the essential element. It takes two H^+ ions for each Ca^{2+} ion but only one H^+ for each

---

BOX 4-1

Chemical Expressions of Elements in Soil Analyses

Relating quantities of elements in terms of mass (weight) as determined in most chemical analyses and relating mass values to reactivity in soils requires a bit of chemistry. The first step is to understand that each atom has a mass. The atomic mass of each element can be found in a periodic table. Atomic mass represents the weight of 1 mole of atoms. A mole is defined as 6.02×10^{23} (Avogadro's number) atoms. Ions are electrically charged atoms or combination of atoms. To relate how much mass of each ion is required to react in chemical reactions, the atomic mass of the ion is divided by the charge on that ion. The equivalent mass of the most common ions competing for one mole of negatively charged CEC sites on clay or organic matter in soil is calculated in the following table.

Most soils only contain between 0.03 and 0.3 mole of charge kg^{-1}. For easier expression in whole numbers, mole charge values are expressed as centimoles (1/100 of a mole) per kilogram of soil (i.e., $cmol_c$ kg^{-1}). (Older soil

Element or Ion	Ion Mass g	Ion Charge	Equivalent Mass g	Equivalent $cmol_c$ mass g
H	1	1+	1	0.01
Na	23	1+	23	0.23
Mg	24	2+	12	0.12
K	39	1+	39	0.39
Ca	40	2+	20	0.20
NH_4	18	1+	18	0.18
Al	27	3+	9	0.09

(continued)

(Continued)

literature expresses charge as milliequivalents per 100 g of soil, i.e., mEq per 100 g of soil.) The numerical values of charge in $cmol_c$ kg^{-1} and mEq $100g^{-1}$ are equal.

From the table (equivalent cmol mass column) we can determine that if a soil has 1 $cmol_c$ kg^{-1} H^+, that is, 0.01 $mole_c$ kg^{-1}, and we want to displace it with Ca^{2+} ions, we need to add 0.2 g of Ca^{2+} ions. To obtain 0.2 g of Ca^{2+} ions from calcium carbonate ($CaCO_3$), we first calculate the molecular weight of $CaCO_3$ by summing $Ca = 40$ g plus $C = 12$ g plus $3 \times O = 48$ g for a value of 100 g. Because each Ca atom in $CaCO_3$ has two charges, the mole equivalent weight of $CaCO_3$ is 50 g. The cmol equivalent weight is 1/100 of the mole equivalent weight or 0.5 g of $CaCO_3$. Thus we would have to add 0.5 g of $CaCO_3$ kg^{-1} to supply enough Ca^{2+} ions to replace 1 $cmol_c$ kg^{-1} of H^+.

BOX 4-2

Cation Exchange Capacity Measurements

There are three basic methods for determining the cation exchange capacity (CEC) of a soil sample. Among different laboratories specific details differ, but the basic difference is the pH value of the sample at which the CEC value is determined. The type of method, that is, the pH used, greatly influences the value obtained in some soil material and has little influence on other soil material. To critically evaluate and compare CEC values, a scientist must know the pH value at which the CEC measurement was made. The three basic CEC methods are as follows:

CEC determination at the pH value of the soil, often called the effective CEC

CEC determination at pH 7, usually called the ammonium acetate method

CEC determination at pH 8.2, often called the sum of cations method

These methods are often abbreviated as ECEC, CEC_7 and $CEC_{8.2}$, respectively.

In ECEC methods, the exchangeable Ca^{2+}, Mg^{2+}, K^+, and Na^+ ions in the sample are displaced by an unbuffered ammonium chloride (NH_4Cl) solution and the Al^{3+} ions are displaced by 1 N KCl solution. The quantity of each element is determined by chemical analysis.

In the $CEC_{7.0}$ methods, the Ca^{2+}, Mg^{2+}, K^+, and Na^+ ions are displaced usually by an ammonium acetate solution buffered at pH 7, and the H^+ and Al^{3+} ions are determined by titration to pH 7.

In the $CEC_{8.2}$ methods, a $BaCl_2$ solution is used to displace the Ca^{2+}, Mg^{2+}, K^+, and Na^+ ions, and titration to pH 8.2 is used to determine H^+ and Al^{3+}.

In each method the centimoles ($cmol_c$) of all the elements determined are summed and divided by the sample weight and expressed as $cmol_c$ kg^{-1} of sample. Older literature expressed the elements in milliequivalents $100g^{-1}$ (mEq $100g^{-1}$) of sample. The two expressions are numerically equal.

Some authors have been careful to indicate the method used, but many are not. As seen in the following table, soil samples

(continued)

(*Continued*)

that are naturally acid and contain large amounts of organic matter and/or kaolinite clay can be expected to have greatly different CEC values related to the pH of the method used. The CEC values of soils rich in montmorillonite clay and containing little or no organic matter are little affected by method.

Soil Sample	Organic Matter %	Soil Sample pH value	ECEC cmol$_c$ kg^{-1} or mEq 100 g^{-1}	CEC$_7$ cmol$_c$ kg^{-1} or mEq 100 g^{-1}	CEC$_{8.2}$ cmol$_c$ kg^{-1} or mEq 100 g^{-1}
Surface soil	14	4.7	15	37	48
Organic soil	50	3.7	16	100	175
Organic soil	50	6.4	137	138	175
Montmorillonite-rich soil	3.5	6.3	50	55	57
Montmorillonite-rich soil	1.0	7.1	42	43	45
Kaolinite-rich soil	6.0	5.0	2	9	22
Kaolinite-rich soil	1.0	5.7	0.2	3	11

SOURCE: Unpublished data from author's laboratory.

K^+ ion. The portion of the CEC sites occupied by Ca^{2+}, Mg^{2+}, K^+, and Na^+ in a soil is known as the base saturation percentage, a valuable indicator of soil fertility (Box 4-3).

Minerals containing aluminum are abundant in soil, and as the concentration of H^+ ions in the soil water increases, aluminum ions (Al^{3+}) dissolve from those soil minerals. As aluminum ions enter into the soil water, they compete for negatively charged sites on the clay and organic matter, further reducing the availability of essential nutrient cations, as shown in Figure 4–3.

The plant-available forms of some essential elements are available to plants as anions (indicated by ⁻ superscripts; see Table 4–2). Anions pass directly into the soil water as organic compounds decompose and inorganic minerals dissolve in the soil. The negatively charged anions are not affected by the negative charges of clay and organic particles (CEC) and tend to move with the water in the soil. Some anions, particularly those of phosphorus, are highly attracted to iron and aluminum oxide minerals, forming very insoluble minerals in the soil, and do not readily move with the flow of water in most soils. Nitrogen as nitrate (NO_3^-) anions and sulfur as (SO_4^{2-}) anions are seldom attracted by minerals in the soil and more readily move in water flowing through the soil.

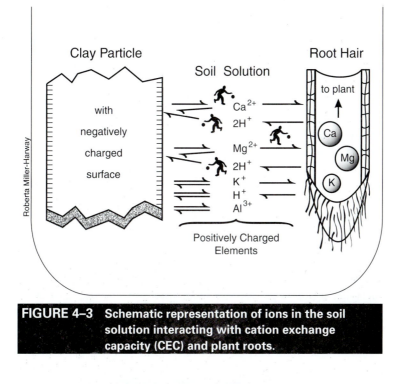

FIGURE 4–3 **Schematic representation of ions in the soil solution interacting with cation exchange capacity (CEC) and plant roots.**

Roberta Miller-Harway

BOX 4-3

Base Saturation Percentage

Base saturation percentage (BS%) is the amount of the CEC occupied by the Ca^{2+}, Mg^{2+}, K^+, and Na^+ ions. These four ions are called base cations. This is the formula for computing base saturation percentage:

$$BS\% = [Ca^{2+} + Mg^{2+} + K^+ + Na^+ \;(cmol_c\;kg^{-1})/CEC\;(cmol_c\;kg^{-1})] \times 100$$

BS% can be expressed on any of the CEC values (i.e., ECEC, CEC_7 or $CEC_{8.2}$ (see Box 4-2)). A base saturation percentage of 50% or greater on a CEC_7 basis is used to define relative fertile soils such as Mollisols (see the appendix). Mollisols must have a BS% of 50% or greater at all depths above 71 inches (1.8 meters). BS% on a $CEC_{8.2}$ base (sum of cations method) measured at a depth of 71 inches (1.8 meters) is used to classify Alfisols and Ultisols (see the appendix). Alfisols must have a BS% value of 35% or greater, and the less fertile Ultisols have lesser values. (Note that slightly different depth criteria apply to specific kinds of Mollisols, Alfisols, and Ultisols.)

THE pH THEME OF SOIL CHEMISTRY

The most significant chemical determinant of an element's availability to plants is the pH value of the water in the soil. A pH value is defined as the negative common logarithm of the H^+ ion concentration in the water. In water, both H^+ and OH^- ions are present. The relative concentration of hydroxyl (OH^-) ions to hydrogen (H^+) ions in water determines the pH value of that water. When the number of OH^- equals the number of H^+ ions, the water is neutral and has a pH value of 7. As the concentration of H^+ ions exceeds that of OH^- ions, the water is increasingly acid and the pH value decreases. Conversely, as the concentration of OH^- exceeds the concentration of H^+ ions, the water is increasingly alkaline (also known as basic) and the pH value increases. The pH values of water in almost all soils are between 10, very strong alkalinity, and 4.0, extreme acidity, but pH values in some soils slightly exceed these ranges. The proportion of the total amount of each essential element in the soil that is available to plants is related to the pH value of the soil water. Availability of some elements is favored by acid conditions and some by alkaline conditions.

In Figure 4–4 the vertical width of the horizontal bar for each element indicates its relative degree of availability throughout a range of soil pH values. The more acid the soil solution (lower pH values), the greater the abundance of H^+ and Al^{3+} ions that compete with the essential Ca^{2+}, Mg^{2+} and K^+ on the clay and organic CEC sites and thereby reduce their availability. However, more hydrogen ions (H^+) in soil solution

FIGURE 4–4 Schematic representation of how pH values influence the availability of the essential elements. The width of each bar represents the relative availability of that element.

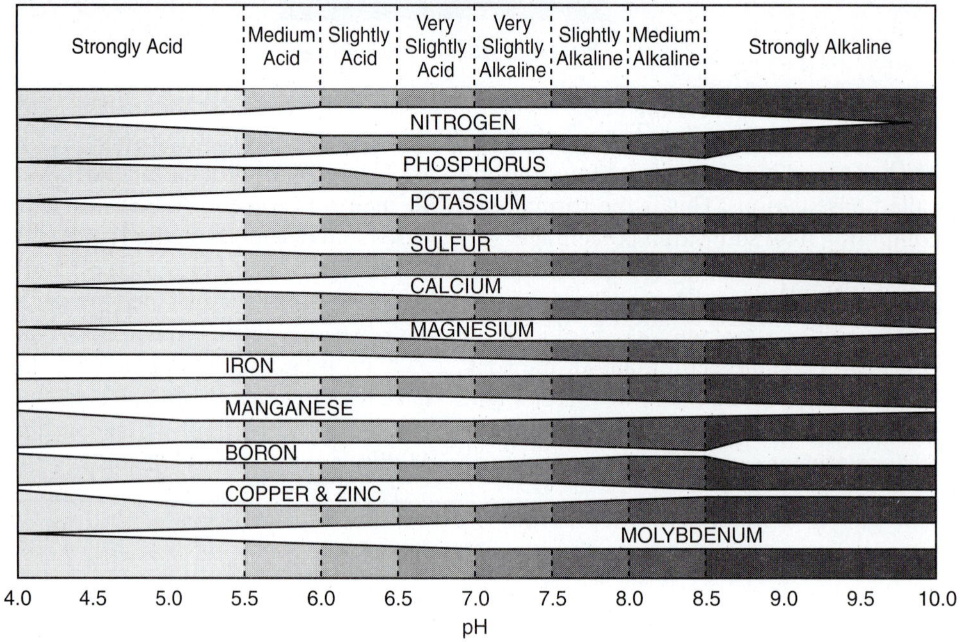

increases the availability of iron, manganese, boron, copper, and zinc by more rapidly releasing them from minerals in the soil. Phosphorus combines in unavailable forms with several minerals in the soil, and the relative availability of phosphorus decreases as the soil water either increases in acidity or alkalinity. A soil pH value near neutral (i.e., pH 7) allows for the greatest plant availability of phosphorus in the soil.

The pH value of the soil is perhaps the most informative chemical evaluation that can be made to estimate availability of essential elements. For greatest availability of all essential elements, a very slightly acid to neutral pH value, between 6.5 and 7.0, is most desirable. To increase the pH value of acid soil, it is a common practice for farmers to add lime, usually as calcium and magnesium carbonate ($Ca:Mg(CO_3)_2$). Adding some form of sulfur that will form sulfuric acid (H_2SO_4) when wetted by water in the soil decreases the pH value of naturally alkaline soils.

ELEMENTAL FORMS NOT DIRECTLY AVAILABLE TO PLANTS

Within the soil all of the essential elements are present in numerous chemical forms not directly available to plant roots. These indirectly available forms of essential elements are often referred to as *unavailable* forms because the essential elements they contain are only available to plants when the structure of the material is destroyed. Unavailable forms of essential elements in the soil can be grouped as either inorganic soil minerals or organic compounds. Both the total amount and the relative proportion of organic compounds and inorganic minerals containing essential elements differ greatly from soil to soil and with depth within the soil. Elements chemically pass from indirectly available, or unavailable, forms to available ionic forms and return to indirectly available forms via several routes. The rate at which the elements move from one form to another depends on the dynamic conditions of temperature, moisture content, and relative abundance of other ions present in the soil, making it impossible to determine exact amounts or proportions of each element available to plants during a growing season.

INORGANIC OR MINERAL FORMS OF ESSENTIAL ELEMENTS

The inorganic forms of essential elements are the silicates, oxides, carbonates, and other mineral components of the sand, silt, and clay particles in the soil. The minerals present in each soil are determined by the composition of the geologic material from which the soil has formed and are known as primary minerals. As these minerals are altered or weathered by conditions present in the soil, new minerals, mostly of clay size, are formed and known as secondary minerals.

The most common primary minerals in soils are *silicates,* which are minerals composed primarily of silica, oxygen, and aluminum. The primary silicate minerals formed at high temperatures of from 1100°F to 2200°F (600°C–1200°C) as molten magma of earth's interior cools some miles beneath the earth's surface. The most common primary silicate mineral in most soils, especially among the sand-size particles, is quartz. Quartz consists of only two elements, silica and oxygen (SiO_2). From the standpoint of soil fertility, quartz provides no essential nutrient elements and can be considered chemically inert as a source of essential elements for life-forms.

Feldspars are primary silicate minerals present in most soils, although quantities are low in many soils. There are several species of feldspar. All feldspars contain silica, aluminum, and oxygen and either sodium, potassium, calcium, or magnesium ions depending on the individual feldspar minerals. Most feldspar minerals are present in soil as sand- or silt-size particles. When exposed to water in the soil feldspars decompose slowly releasing their ions into the soil water. As the feldspars decompose, potassium (K^+), calcium (Ca^{2+}), and magnesium (Mg^{2+}) ions are released into the soil water and are attracted to negative CEC sites on the clay and organic particles in the soil. These essential elements are required in rather large amounts for plant growth (Table 4–1). The rate at which these elements are released by the weathering of primary minerals is seldom sufficient to supply the needs of rapidly growing food crops but may be rapid enough to sustain slow-growing forest vegetation.

Micas are primary silicate minerals with flat platelike particles that are frequently seen to sparkle when present as sand-size particles in soil (Figure 2–7). Like feldspars, micas form at high temperatures and are primarily composed of aluminum, silica, and oxygen, but most important to soil fertility they contain potassium. In the soil, primary mica particles tend to break physically into particles of sand, silt, and clay sizes. Along with some of the feldspars, micas are a source of potassium ions as they decompose. As micas decompose and release potassium ions, the secondary mineral vermiculite is formed. Vermiculite has the ability recapture K^+ and also NH_4^+ ions from the soil solution when the soil becomes dry. Some mica minerals (e.g., biotite) also contain significant amounts of iron, and as they decompose in the environment of the soil, the iron combines with oxygen to form red-colored iron oxides in the soil.

Almost all soils have a mixture of primary silicate minerals as sand- and silt-sized particles and secondary clay-sized silicate and oxide minerals. As primary silicate minerals slowly decompose (weather) in the soil, the silicon, aluminum, hydrogen, and oxygen ions that are released reform into secondary silicate minerals and oxides of clay size. Most clay-size silicate minerals are platelike. Kaolinite is the simplest clay, containing only silica, aluminum, oxygen, and hydrogen and no essential elements. Kaolinite clay tends to form as primary minerals weather in acidic environments. Kaolinite has relatively few negative charges (CEC) per unit of weight. Kaolinite does not expand when wet nor shrinks when dry. Halloysite has a chemical composition like kaolinite but has a tubular form.

Montmorillonite is a clay mineral that also forms from the decomposition of feldspars and micas. Montmorillonite formation is favored in alkaline environments where the water has a high concentration of magnesium. Montmorillonite expands upon wetting and becomes very sticky. Its presence in high quantities is responsible for extreme soil cracking when a soil becomes dry. Soils with high contents of clay-size montmorillonite are notorious for causing house foundations and roadbeds to fail because of soil expansion upon wetting and contraction upon drying. Montmorillonite has roughly ten times more negative charges (CEC) per unit weight than the other clays and therefore can hold a proportionately greater amount of positively charged elements available for roots. However, in acid conditions, aluminum (Al^{3+}) or hydrogen (H^+) ions are attracted to the negatively charged CEC sites on the clay, and only small amounts of potassium (K^+), calcium (Ca^{2+}), and magnesium (Mg^{2+}) ions are present. Large amounts of lime are required to raise the pH of acid soils containing appreciable amounts of montmorillonite. The high amount of negative charge (CEC) on montmorillonite clay is effective in capturing many potentially toxic substances that otherwise may leach through the soil into the groundwater. When wet and expanded, water movement through soils with a high content of montmorillonite clay is very slow. Montmorillonite clay is frequently used to line the bottom and sides of waste disposal sites.

Oxides in the soil are primarily compounds of iron (Fe_2O_3), aluminum (Al_2O_3), or manganese (Mn_3O_4). Each of these oxides is present in several oxide forms. Most often the oxides are present as clay-size particles, but they may aggregate into larger particles or act as glue aggregating other soil particles into sand- and gravel-sized concretions and nodules. Iron oxides tend to coat the surfaces of the silicate minerals, and their presence is easily detected by the red and yellow colors these coatings impart to the soil. Manganese oxides are black; aluminum oxide is gray or off white. These oxides form when the metal ions released during the weathering and decomposition of iron-bearing silicates combine with oxygen. If soil is saturated with water, little or no oxygen is available, and the iron and manganese oxides become soluble and leach from the soil. The absence of iron oxide to coat the silicate mineral particles is indicated by gray-colored soil.

Iron and aluminum often react with the available forms of phosphorus to form the very inert minerals strengite ($FePO_4$) and variscite ($AlPO_4$), respectively. These reactions reduce the plant availability of P in the soil and are commonly referred to as P fixation reactions. Their formation also retains phosphorus from leaching into the groundwater.

Carbonates are minerals formed as water evaporates to the atmosphere leaving behind the elements that were dissolved in that water. The most common carbonate minerals in soils are calcite and dolomite. Calcite is a calcium carbonate ($CaCO_3$) mineral, and dolomite is a combination of calcium and magnesium carbonate ($Ca:Mg(CO_3)_2$). The most concentrated forms of carbonates are present in limestone rocks. Carbonates are rather soluble in acidic water, and in humid areas they dissolve and leach from the soil. In less humid areas, carbonates often become concentrated a few inches to a few feet below the soil surface as infiltrating water dissolves them in the surface layers of the soil, and they precipitate in the subsoil as water is transpired by growing plants. Carbonates are important sources of calcium and magnesium in soils. Limestone forms of carbonates are routinely pulverized into small particles and added as liming material to increase the pH value of acid soil.

Numerous other minerals are present in lesser quantities in most soils. Of these, the most significant in the role of soil fertility are the phosphorus-bearing minerals. Phosphorus-bearing minerals are the *only* primary sources of phosphate for plant and animal life. Note in Table 4–1 that phosphorous is more concentrated in plants and humans than in the soil. It is the only major element required for life whose natural source is soil minerals for which this is true. Phosphate in soils is a component of the iron and aluminum phosphate minerals strengite and variscite and a group of calcium phosphate minerals known as apatites. Apatite minerals have numerous chemical variations, but the general form ($Ca_5(PO_4)_3(OH,F)$), with various OH and F concentrations, is probably the most common. In most soil conditions, calcium phosphates are much more soluble and provide higher concentrations of plant-available phosphorus than the iron and aluminum phosphate minerals. Apatites are most concentrated in soils formed from limestone and other calcium-rich sedimentary rocks that were at one time at the bottom of shallow seas. Most marine sediments contain the skeletons of sea animals, rich in phosphorus and calcium, which were deposited over millions of years and lifted out of the sea by movements of the earth's crust.

Soils that have above-normal supplies of rather soluble phosphate minerals are well known as excellent habitat for racing horses. The bluegrass area around Lexington, Kentucky, USA, is one such area. The abundance of phosphate in the soil formed from phosphorus-rich limestone rocks apparently leads to stronger than average bone and muscle structure in the horses that consume the grass grown on such soil.

ORGANIC FORMS OF ESSENTIAL ELEMENTS IN THE SOIL

Organic compounds containing essential elements enter into in the soil only after they have been part of a living organism. We regularly see the most common process of introducing organic forms of the essential elements into the soil as leaves and other dead plant parts are deposited on the soil surface. All organic compounds found in soil were formed as cells in organisms living both in the soil and on the land surface. The major components of organic compounds are C, H, and O, the plants obtained from the air. Because organic compounds were formed in living cells, they also contain all of the other elements necessary for life that plants and microbes obtained from the minerals in the soil. When cells die and tissues decompose, their elemental components are released as inorganic ions in the soil water and become sources of essential elements for actively growing plants.

Although nonliving organic compounds in the soil are too numerous to identify, they can be conceptualized as being in two groups. During the early stages of decomposition, nonliving organic tissues are identified as *organic residues,* which are plant and animal tissues that retain identifiable structure. A person can identify the organism or part of the organism from which the organic material originated as a leaf, a bone, excrement from an animal, and so on.

After several stages of decomposition, the nonliving organic material loses all structural resemblance to its living origin and becomes an unrecognizable organic material called *humus.* Although structural form is not recognizable, humus imparts black color to soil. Very small particles of humus coat the mineral particles in the soil, and in the presence of only a small amount of humus material the entire soil mass will have a degree of black color.

The amount of organic compounds in the soil is most often referred to as soil organic matter (SOM) content. In almost all analyses of SOM, plant and animal tissue fragments larger than 0.079 inches (2 mm) in diameter are routinely removed by sieving the soil sample before analysis. This removes most organic residue. After removing organic residue from a soil sample, the organic carbon content of the soil is chemically determined. Soil organic carbon (SOC) content is usually multiplied by a factor of 1.72 to express a SOM content of the sample.[1] Except for very detailed studies, SOM content data only identifies the amount of organic carbon in the soil and provides no quantifiable information about the form or amount of essential nutrients present.

Organic carbon additions to soil are greatest near the soil surface because organic residues are physically mixed into the mineral soil. The black color of organic carbon compounds is often associated with the popular term *topsoil.* Near the soil surface, plant roots, insects, worms, and microorganisms living in the soil also contribute organic tissue upon death. Decomposition and oxidation of the plant and animal tissues begins as soon as the plant dies if temperature and moisture are suitable for microbiological activity. During decomposition a high proportion of the organic carbon contained in the organic matter is combined with oxygen by the microbes to form CO_2 gas and returns to the air. Hydrogen ions (H^+) from the decomposing organic molecules acidify the soil water. The soil microbes incorporate some of the carbon and other essential elements into a multitude of new organic compounds within their cells. However, soil microbes also die and contribute their cells as nonliving organic material for another generation of microbes to decompose.

Plants are not able to ingest significant amounts of essential nutrient elements while they are chemically attached to most organic molecules. However, during each stage of decomposition, essential elements are released from the decomposing organic molecules and enter the soil solution as *inorganic* ions at which time they are available for both plant roots and microbes in the soil. Many variables control the rate of elemental release from organic compounds in the soil. The most rapid release is from the decomposition of recently deposited plant and animal remains. The humus formed during successive stages of decomposition appears to be more resistant to decomposition.

Although some organic carbon may persist in soil as humus for several hundred or even thousands of years, almost all of the organic carbon eventually is released from the soil as CO_2 gas and returns to the air. If conditions of vegetative density, temperature, and moisture are maintained, each soil reaches a *steady state* where the rate of organic carbon created by growing plants and deposited in or on the soil as nonliving organic residue equals the rate at which organic compounds are decomposed to CO_2 gas. At steady state the amount of organic carbon in the soil remains nearly constant over time.

Colder soils usually have higher contents of organic carbon than warmer soils. Soils that are saturated with water usually contain more organic carbon and have darker colored topsoil than soils that are usually dry. This appears related not only to the fact that the oxygen supply in the soil may be limited in saturated soil but also to the fact that wet soils have a higher heat capacity and do not become as hot during daytime heating by the sun as dry soils. This same effect can be used to explain why soils under the shade of trees contain more organic carbon than cultivated soils that are exposed to direct sunshine much of the year. Clearing natural vegetation in preparation for cultivated crops almost always causes reduction in soil organic carbon content. With continued cultivation a new steady state of soil organic will develop. When arid soils are irrigated for crop production soil organic matter contents often increase. Although clear evidence indicates that soils in colder climates contain more organic carbon than soils in warm climates, it now appears that maximum daytime temperatures may be a primary factor in SOC content. Modern data clearly show that when well-drained mineral soils with the same mean annual soil temperature from the tropical and temperate latitudes are compared, soils in the tropics have higher organic carbon contents than soils in the temperate areas (Figure 4–5). These observations can be explained by remembering that to have equal mean annual soil temperature, soil temperatures during the summer in the temperate latitudes have to be much higher than any experienced in tropical areas where temperatures remain relatively constant throughout the year. Also, in the tropics, temperatures are warm throughout the year so when water is available, plants

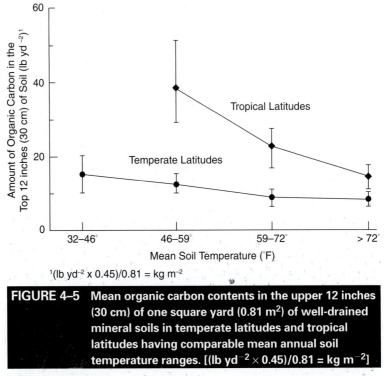

^{1}(lb yd^{-2} x 0.45)/0.81 = kg m^{-2}

FIGURE 4–5 **Mean organic carbon contents in the upper 12 inches (30 cm) of one square yard (0.81 m^{2}) of well-drained mineral soils in temperate latitudes and tropical latitudes having comparable mean annual soil temperature ranges. [(lb yd^{-2} × 0.45)/0.81 = kg m^{-2}]**

Source: Buol et al., 1990.

fix carbon from the air and produce more organic residue than in temperate latitudes where plant growth slows or ceases in the winter season.

In soils saturated with water much of the year, no voids are available for air. Under these conditions the supply of oxygen in the soil limits microbial ability to decompose organic compounds. Where saturated conditions are present for long periods each year, SOM accumulates to form the organic soils commonly known as peat or muck and taxonomically designated as Histosols (see the appendix). These soils have black color and high organic carbon contents but are not always fertile. Some Histosols are extremely acid and chemically unsatisfactory for the growth of crop plants (see Chapter 6).

The amount of essential elements in organic material varies greatly depending on the chemical composition of the organism from which the material was derived. Some carbon compounds from plants grown on soils with low contents of essential elements have only small quantities of essential elements relative to the content of carbon. Most organic residues with low contents of essential elements relative to the content of carbon decompose more slowly than nutrient-rich organic residues. It is the rate at which decomposing organic material releases essential elements into the soil water as inorganic ions that are available to plant roots, not the amount of organic carbon present in the soil that is vital to essential element availability from organic materials. Heating the soil greatly increases the rate at which SOM decomposes and releases essential nutrient elements it contains. Removal of vegetation that shades the soil exposes the soil to direct sunlight and higher soil temperatures, which speeds the process of oxidizing the carbon in SOM to CO_2 and the release of essential elements as inorganic ions. The burning of organic residues greatly speeds the release of essential elements.

FORMS OF MAJOR ESSENTIAL ELEMENTS IN SOIL

Each essential element has unique chemical properties. Therefore it is necessary to examine briefly how and under what conditions the plant-available forms of each essential element in Table 4–2 interacts with mineral and organic materials in the soil.

Nitrogen

Almost all of the nitrogen in the soil is associated with organic compounds. A few nitrate minerals are present in some soils in arid regions, but nitrate minerals are quite soluble in water and are seldom present in humid areas. Small amounts of nitrogen enter the soil in rainwater. Strong electrical discharges in thunderstorms convert some N_2 gas in the air to nitrate (NO_3^-) and ammonium (NH_4^+) that is carried to the ground in precipitation. It is estimated that about 10 pounds of nitrogen per acre (11 kg ha^{-1}) per year are added to a soil in rainfall, although amounts differ by location. This nitrogen enters the soil solution, where it is readily available to both plants and soil microbes.

Nitrogen (N_2) from the air is converted to available forms primarily through the activity of microorganisms present in a symbiotic association with higher plants (mainly *Rhizobia* species) and nonsymbiotic microorganisms that are independent of higher plants (mainly *Azotobacter, Clostridium,* and *blue green algae* species) in the soil. The symbiotic bacteria live only with higher plants known as legumes, the most common of which are alfalfa, clover, and beans. The rate at which symbiotic bacteria can convert N_2 gas from the air into organically bound nitrogen is known to reach over 300 pounds per acre (336 kg ha^{-1}) per year, although half that amount is more usual. The nonsymbiotic microorganisms present in soil regardless of plant vegetation may convert approximately 10 to 20 pounds of nitrogen per acre (11–22 kg ha^{-1}) per year under some conditions.

Nitrogen taken from the air by nitrogen-fixing microorganisms is incorporated into their biomass forming a multitude of organic carbon-nitrogen compounds. When the microorganisms die and the carbon in their cells is oxidized to CO_2, nitrogen is released as inorganic nitrate (NO_3^-) and ammonium (NH_4^+) ions in the soil water. The process of converting organic forms of nitrogen into plant-available inorganic forms is known as nitrogen *mineralization*. A soil pH value near 7 favors microbial activity that mineralizes organic materials, accounting for greater N availability at pH values near 7 as illustrated in Figure 4–4.

Microorganisms need to maintain a ratio of carbon to nitrogen (C:N ratio) of about 10:1 in their cells. Microorganisms, being extremely numerous and active in warm moist soil, are so aggressive in obtaining nitrogen for their metabolism that plant roots may not obtain much nitrogen from the decomposition of organic matter until the C:N ratio of organic compounds in the soil is less than about 32:1. In Table 4–3 we see that several plant parts, especially wood, have far higher C:N ratios. The stalks and leaves of common grain crops have C:N ratios of 60:1 and 80:1. Microbes in the soil

TABLE 4–3 Approximate C:N Ratios of Various Organic Residues	
Organic Residue	*Representative C:N Ratio*
Tree wood	600:1
Tree bark	450:1
Oat straw	80:1
Cornstalks	60:1
Alfalfa hay	13:1
Animal feces	9:1

decompose these materials, exhausting the carbon as CO_2, but they incorporate all the nitrogen released into their cells until the C:N ratio of the decomposing organic residues is about 32:1. This may take from 8 to 15 weeks, depending on temperature and moisture conditions for straws and much longer for woody residues. During that time, plants growing in the soil to which organic materials with a high C:N ratio has been added, will suffer from lack of nitrogen because the microbes are taking available NO_3^- and NH_4^+ from the soil solution. Only organic residues from legumes such as alfalfa and animal feces provide available N immediately upon initial decomposition in the soil.

Although both ammonium (NH_4^+) and nitrate (NO_3^-) can be present in the soil solution, most nitrogen is present as NO_3^-. Although NH_4^+ can be held as exchangeable ions by the negative charges on the clay and organic matter particles, the NO_3^- is subject to leaching if excess water moves through the soil. Nitrate (NO_3^-) can also be lost from the soil solution if there is a lack of O_2 dissolved in the soil water. This happens when the soil is saturated with water. When a source of O_2 from air or dissolved in the water is not available, some species of microbes in the soil have the ability to take O from NO_3^- molecules, in which case NO_3^- is converted into volatile N_2, NO, or NO_2 gas and escapes back to the air. This process is known as *denitrification.*

With essentially no source of nitrogen in the mineral components of soil, the amount of nitrogen that can be supplied to a crop is limited to small amounts of N dissolved in precipitation plus amounts that can be fixed as organic compounds by N-fixing microbes, both symbiotic and nonsymbiotic species. In addition to nitrogen removed by harvested crops, nitrogen is subject to removal from weather events that cause nitrate leaching and denitrification. The only reserve of nitrogen contained in soil is SOM, preferably with a C:N ratio of less than 32:1. For plants to access the nitrogen contained in SOM, the organic carbon must be oxidized to CO_2 and exhausted to the air; that is, SOM content decreases.

Phosphorus

Phosphorus (P) originates only in inorganic minerals of the earth's crust. Plants must have phosphorus for metabolic functions and as a component of cell nuclei. Phosphorus deficiency leads to softening of bones in adult animals (osteomalacia). Phosphorous must be in a soluble ionic ($H_2PO_4^-$ or HPO_4^{2-}) form for use by plant roots. There are many mineral forms of phosphorus in primary soil minerals. Each phosphorus-bearing mineral has its own susceptibility to dissolution. The calcium apatites are the most soluble; P contained in iron and aluminum oxides is much less soluble, except under certain conditions.

Acidity or alkalinity (i.e., pH) of the soil is critical for maintaining plant-available forms of phosphorus in the soil (Figure 4–4). More of the phosphorus in the soil is attracted to insoluble iron and aluminum oxide forms in soil with acid pH values below 5.5. When soil pH values exceed 7.5, the calcium forms of phosphorus are less soluble. At pH values above 8.5, highly soluble forms of sodium phosphates are present, but this is a rare occurrence in soils and of little practical significance.

The relative abundance of aluminum and iron oxides in most soils assures that soluble forms of phosphorus seldom escape the soil via leaching. Phosphorus is most likely to leach in saturated sandy soils with low iron oxide content or in organic soils. Phosphorus is lost when solid soil particles are removed by erosion. The major loss of phosphorus from soil used for food crop production is via removal in the harvested crop. High yields of grain crops extract between 10 and 30 pounds of phosphorus per acre (11–33 kg ha^{-1}) per crop.

Of course, phosphorus is present in the organic residue of dead plants and animals, especially animal bones. Crushed animal bones were known to improve plant growth and used as fertilizer by early civilizations. All organic forms of phosphorus are secondary and contain only phosphorus that plants initially obtained from primary soil mineral sources. As organic materials decompose, the phosphorus is released into soluble forms. More resistant forms of SOM also contain phosphorus, but because they resist decomposition they only slowly provide plant-available phosphorus. Like the C:N ratio, the C:P ratio in organic material is important in determining how much P will become available as the organic material decomposes. It appears that a C:P ratio of less than about 60:1 is required if P released from SOM decomposition is going to become beneficial to crop plants. The microbes in the soil capture most of the released phosphorus when organic materials with C:P ratios greater than 60:1 decompose.

Usually the surface layers of soils contain larger quantities of phosphorus than subsoil layers. Under undisturbed vegetation, plants concentrate phosphorus and other essential elements obtained from their entire rooting depth and deposit them as organic residue on the soil surface as they die.

No absolute value can be given for what proportion of the total amount of phosphorus present in a soil is available to crop plants during the course of a growing season, but an average value of 1% is a reasonable approximation. Unlike nitrogen, there is no recharge of phosphate possible from the air, except minute quantities in dust, so clearly there is only a finite amount of phosphorus, determined primarily by the mineral composition of the soil available for export via harvested food crops.

Potassium

Potassium originates only in minerals within rocks and soils. Plants require potassium in rather large amounts primarily as an integral part of their metabolism in the formation and transport of carbohydrates and proteins. Unlike phosphorus, which is concentrated in the seed, a majority of the potassium is found in the stems and leaves of plants.

The principal minerals containing potassium in soil are micas and feldspars. It is not possible to be specific about what proportion of the total potassium in the soil is available for plant uptake, but about 1% is a reasonable value for a conceptual understanding of potassium fertility. The available form of potassium is the potassium cation (K^+).

In the soil, potassium cations (K^+) are released as silicate minerals decompose and become attracted to the negatively charged surfaces of clay minerals and organic compounds. They are referred to as *exchangeable* potassium (Figure 4–3). Potassium ions (K^+) are less tightly attracted to these negatively charged sites than hydrogen (H^+) or aluminum (Al^{3+}); thus in acid soils, or sandy soils with little negative charge, potassium is somewhat subject to leaching. Also, an oversupply of calcium (Ca^{2+}) can displace the K^+ ions and decrease the availability of potassium if it exceeds potassium by more than 100:1.

The most potassium-deficient soils are those that contain little or no mica or feldspar minerals. Often these are sandy deposits where the micas and feldspars have been decomposed prior to deposition. Also, soils formed from sandstones are frequently low in potassium content because the sandstone is composed mainly of quartz (SiO_2). The feldspar and mica minerals, if ever present, were dissolved during the geologic millennia when the sandstones were deposited.

Unlike the other essential elements, potassium is quickly released from organic residues on the soil's surface. Potassium is present in the most easily decomposed parts of plant cells, and it rapidly leaches from organic residue soon after a tissue dies. Like phosphorus, the amount of potassium in a soil is determined by the finite quantity present in the soil minerals, and the quantities of potassium removed by crop harvest are of such a magnitude that sustainable crop production must involve the replenishment of potassium harvested in the food products.

THE OTHER ESSENTIAL ELEMENTS

Other plant-essential elements each have specific chemistries, but most follow a pattern similar to either phosphorus or potassium. *Sulfur* is an exception because it is frequently supplied from aerial deposition, especially in industrialized areas. Significant amounts of sulfur in the air can be traced to the burning of coal and is returned to the soil via acid rain. In air relatively free of industrial smoke, crop plant growth is frequently limited by a lack of sulfur. With societal efforts to reduce sulfur emissions, it has been necessary to fertilize crops with sulfur in some areas. The beneficial effects of adding sulfur are frequently noted in remote areas of the tropics where the air is relatively free of sulfur. Sulfur is a component of several commercial fertilizers, and adequate quantities of sulfur are often derived as a by-product with other fertilizers.

Calcium and magnesium are available to plants as exchangeable divalent cations, Ca^{2+} and Mg^{2+}, respectively. They, like K^+, are retained in the soil on negatively charged clay and organic matter (Figure 4–3). A substantial amount of calcium is required as a structural component of cell walls in plants. Magnesium is essential to chlorophyll formation and photosynthesis. Calcium and magnesium are contained in many soil minerals but concentrated in limestone rock. Deficiency of these elements only occurs in acid soils. Adding crushed limestone to increase soil pH values is a routine practice in modern agricultural management of acid soils, and deficiencies of Ca and Mg are rare in sustained agricultural production but may occur in very acid soil.

Being a component of cell walls, calcium is not translocated within a plant. Some exchangeable calcium must be present at that point in the soil where the plant root tip is growing. Many virgin soils in humid areas contain very little exchangeable calcium (Ca^{2+}) in their subsoil. In such soils plant roots do not enter the subsoil except along channels created by insects or animals that burrow and physically incorporate calcium-bearing material or deposit some calcium contained in their cells as they die. Calcium and magnesium move downward slowly in the soil because they are retained by the negative charges of clay and organic matter. Physical mixing of liming material deep into the soil is usually required for best results. Soils that have a very low amount of negative charges (CEC) permit some calcium and magnesium movement in percolating water, and surface applications of soluble calcium materials such as gypsum ($CaSO_4 \cdot 2H_2O$) can be used to enrich the subsoil with calcium and thereby promote increased root growth. It may take several years, but the calcium enrichment of naturally acid subsoil is clearly evident after many years of lime applications (Buol and Stokes, 1997). Soils formed from limestone and most soils in arid regions have abundant quantities of calcium and magnesium in the subsoil.

Iron is a component of several soil minerals. Iron oxides, if not masked by the black color of SOM, give soil a yellow or red color. Available iron is almost always present in more than adequate amounts; however, when the soil pH is too high, usually above 7.5, iron compounds become so insoluble that plant growth may be restricted by lack of soluble iron. The leaves of plants that are deficient in iron characteristically have a light green or yellowish green color.

When soil is saturated for prolonged periods of time, anaerobic conditions develop as microorganisms in the soil use oxygen in their metabolism and air containing oxygen is prevented from entering the soil because all the voids are filled with water. Under these conditions any iron that is present or released from iron-bearing minerals is chemically reduced to the ferrous form (Fe^{2+}), becomes soluble, and leaches. With no iron to precipitate on the surface of the silicate minerals or as separate particles, the entire soil retains the gray color of the silicate minerals. Gray-colored soils have small contents of iron oxide, but iron is seldom limiting to plant fertility except at the high pH values just mentioned.

Manganese is present in most soils but may be lacking in some soils formed under waterlogged, anaerobic conditions where the soluble forms have been removed. Also, acid soil conditions may solubilize so much manganese that it becomes toxic to certain plant roots. Maintaining a soil pH value between 5.2 and 7.5 is desirable for manganese fertility.

Zinc, copper, boron, and *molybdenum* deficiencies may occur in some soils and under certain conditions. Often deficiencies are regional, related to a lack of the element in the geologic material from which the soils formed. Because plants require only small quantities of each of these elements and solubility is related to soil pH (Figure 4–4) and moisture content, their deficiency may be transient with weather conditions. Higher soil temperatures increase microbial activity and the release of these elements from organic compounds. At lower soil temperatures, organic matter oxidation is slow and crops may experience deficiency not observed when the soil is warmer and the rate of organic matter oxidation is more rapid.

Iron, manganese, zinc, copper, boron, and molybdenum are often referred to as *micro* or *trace* elements. The small quantities required by plants can sometimes be supplied to crops by foliar application of fertilizers. Applying soluble iron compounds as foliar sprays is a rather common practice to attain dark green color in pine trees being readied for Christmas tree sales. Citrus trees growing in soil with high pH values are frequent victims of iron deficiency. Driving an iron nail or two into the trunk of citrus trees has been reported to relieve iron-deficient conditions effectively.

DETERMINING THE FERTILITY OF A SOIL

Soil-testing methods for determining the relative amount of available nutrients in soils have been researched for many years. The goal of soil testing is to predict the probability of a profitable response to applications of fertilizer and lime. Soils that test low in a particular element have a high probability of a profitable response to applications of that element; soils testing high have a low probability of a profitable response to fertilizer application. A large number of soil-testing procedures have been developed over the years and are routinely used to recommend rates of fertilizer and lime on commercial crops. Soil-testing techniques usually extract a soil sample with a dilute acid that removes exchangeable ions and attacks some organic compounds and some soil minerals in an attempt to determine a comparative index of phosphorus, potassium, calcium, and magnesium and in some instances certain micronutrients that will be available during the crop-growing season. Soil-testing techniques are very good where experienced scientists apply the technology to established methods of farming and a known range of soil conditions. A single technique for soil testing that accurately predicts available nutrient supplies in all soils and for all types of plants is probably unattainable. There are simply too many variables. Among the most confounding problems are the wide range of inorganic minerals and organic compounds that exist in different soils, root characteristics that differ from plant species to plant species, and contrasting weather patterns from year to year. Most soil-testing methods serve well to predict the fertilizer phosphorus and potassium requirements of most common crops because their chemical interactions are largely controlled by inorganic reactions.

Diagnosis of nitrogen supplies is more problematic because microbial activity and organic reactions that control the release of nitrogen from organic compounds are very dependent on unpredictable moisture and temperature dynamics in the soil. Soil-testing techniques that have evolved to predict the amounts of fertilizer needed to supply nutrient elements for major agricultural crop species often have limited applicability to natural ecosystems where nutrient demand is relatively low compared to that of food crops.

PERSPECTIVE ON THE ESSENTIAL ELEMENTS AND FERTILITY OF SOILS

An adequate supply of *each* essential element is necessary for good plant growth (i.e., *the law of the minimum*). Although carbon, oxygen, and hydrogen are the most abundant elements in plant and animal tissue, they are supplied by air and water, and a deficiency of these elements is seldom of concern in food production. Increased amounts of CO_2 in the air are known to increase the growth rate of many plants, but except where CO_2 content in the air can be controlled, as in greenhouses, it is not a viable management option.

Nitrogen, phosphorus, and potassium are the major elements of concern in soil fertility. Inadequate supplies of one or more of these elements are most often found to limit the growth of food crops because of the relatively large amounts of these elements removed as plants are harvested in food crop production. A deficiency of any of the other essential elements, often referred to as trace or micro elements, can be just as limiting to plant growth. Detection of micro nutrient deficiencies via soil tests is problematic because of the small quantities required by plants. However, when a micro nutrient deficiency is detected, only small amounts of the proper kind of fertilizer are needed.

The availability of essential elements is closely related to soil pH values. Increasing acidity is a constant process as H^+ are released from growing roots in all but the most lime-rich soils. Determining soil pH value and correcting acidic conditions with amendments of lime (calcium and magnesium carbonate) is an almost universal requirement for sustained food crop production in most humid areas. Correcting pH conditions that are too alkaline for optimum nutrient availability is required in some soils, mainly in arid areas.

The total amount of each essential element present in the soil has little or no relationship to how much of each element will become available to growing plants. The availability of essential elements is controlled by interactions among the multitude of minerals and organic compounds that are different in both quality and quantity from soil to soil. Where soil composition is known, modern soil-testing techniques interpreted by skilled scientists have proven successful for predicting fertilizer requirements of major food crops that have a large demand for nutrients. Unpredictable weather variables of moisture and temperature during a specific growing season complicate even the most sophisticated methods of essential element availability assessment.

LITERATURE CITED

Buol, S. W., P. A. Sanchez, J. M. Kimble, and S. B. Weed. 1990. Predicted Impact of Climatic Warming on Soil Properties and Use. In B. A. Kimball, M. J. Rosenberg, and L. H. Allen, Jr. (Eds.), *Impact of Carbon Dioxide, Trace Gases, and Climate Change on Global Agriculture* (pp. 71–82). American Society of Agronomy Special Pub. No. 53. Madison, WI.

Buol, S. W., and M. L. Stokes. 1997. Soil Profile Alteration Under Long-Term, High-Input Agriculture. In R. J. Busesh, P. A. Sanchez, and F. Calhoun (Eds.), *Replenishing Soil Fertility in Africa* (pp. 97–109). Soil Science Society of America Special Pub. No. 51. Madison, WI.

NOTE

1. The 1.72 conversion factor is known to be in error because there are so many organic compounds in SOM. Although SOM content is commonly reported, organic carbon content (SOC) is a more correct expression.

CHAPTER REVIEW PROBLEMS

1. What material most often is responsible for black color in soil?
2. Name the three elements that make up most of the weight of plants. From where do plants obtain these elements?
3. What processes have to take place before nitrogen taken from the air by nitrogen-fixing bacteria is available to nonlegume plants?
4. How does pH value of a soil influence the availability of phosphorus in the soil?
5. Which of the essential elements have their availability decreased by high pH values in the soil? Which essential elements have their availability decreased by low pH values in the soil?
6. Which essential element composes more than 15% of the human body but less than 1% of soil material?
7. Which essential element derived from the soil is ten times more concentrated in the human body than in the soil?
8. Which essential element is derived from the air but must first pass through the soil before it can be taken up, in large quantities, by the plant?

5

Elemental Transfers in the Soil, Land, and Life System

All naturally occurring chemical elements are present in soil. Exact identification of all chemical compounds in soil is not possible. A detailed accounting of known compounds and minerals in soil would fill several volumes. In Chapter 4 the life-essential elements were identified and related to various chemical forms in which they are present in air, soil, and living organisms. In this chapter the form and distribution of elements within soil are represented as three functional "pools." The functional pools in soil are identified as *available, organic,* and *mineral* (Buol, 1995). Figure 5–1 shows a schematic diagram of the chemical and physical movement of the essential elements to and from the pools within the soil and interconnection to life and human activity above the land.

Plants powered by radiation from the sun intercept essential elements as inorganic ions from the available pool and combine them with carbon from the air and hydrogen and oxygen from water and air to form carbohydrates, sugars, proteins, fat, vitamins, and other organic compounds. If the plant is not removed, these elements are deposited on the soil surface when each plant dies. This is not the case in the production of human food. At the top of Figure 5–1 human interventions in the movement of essential elements are depicted as a removal of essential elements from soil, land, and life systems via the movement of food to cities. The term *city* includes even small villages and connotes a central site where humans consume food and dispose of waste materials.

The concept of nutrient pools in the soil illustrates the unseen transformations that take place as essential elements are cycled between generations of life-forms. Water is not portrayed in Figure 5–1 but *must* be present in required amounts for almost all of the transformations to take place. *Energy* in the form of radiation from the sun and temperatures compatible with plant life also *must* be present. The interchange of energy, water, and essential elements necessary for plant growth takes place on the surface of the land, in plants, and in the soil immediately below. Water and temperature, as discussed in Chapters 2 and 3, respectively, are variable and dynamic conditions in the soils at the land surfaces of the earth. The flow of essential elements to the plant roots is also variable and dynamic, as discussed in Chapter 4. Before we follow the elements as they pass through the soil and plant systems, let's take a look at each pool and how elements pass into and out of each pool.

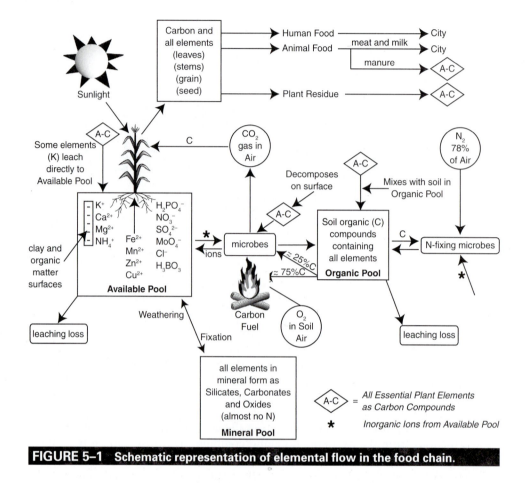

FIGURE 5–1 Schematic representation of elemental flow in the food chain.

MINERAL POOL

All of the essential elements for which soil is identified as the natural source in Table 4–1 originate in the mineral pool, which consists of minerals of geologic origin. Although altered to some extent by weathering processes in soil, the essential elemental composition of the mineral pool is related to rock type. A brief overview of various rock types, their origin, and relative essential element content provides a basis to understand the natural chemical fertility of soils formed from various geologic materials.

Blatt and Jones (1975) have reported the relative proportion of rock types from which soils have formed on the land surfaces of the earth. *Crystalline* rocks are formed by the cooling of the earth's magma and compose only 34% of the present land surface. Crystalline rocks are divided into three major groups. *Intrusive crystalline* rocks are present under 9% of the earth's land surface. They formed from slow-cooling magma deep below the earth's surface and thus are composed of large mineral crystals. Granite is a common crystalline rock. Phosphorus and calcium contents tend to be relatively low in soils formed from these materials. When exposed at the land surface, intrusive crystalline rocks tend to weather slowly and form relatively infertile soils with sandy

FIGURE 5–2 Photo of a recent lava flow, hardened to basalt on the island of Hawaii, Hawaii, USA.

surfaces and clayey textured subsoil. *Extrusive crystalline* rocks are formed by rapid cooling of molten magma deposited on the land surface by volcanic activity and are present under 8% of the land area. Basalt is a common extrusive rock (Figure 5–2). Extrusive rocks have small mineral grain size and most form relatively fertile clayey and loamy soils. *Metamorphic rocks* in most cases have a mineral composition like intrusive crystalline rocks but have been physically altered by extreme temperatures and pressures in the earth's crust and have mineral grain sizes intermediate between the intrusive and extrusive rocks. Upon weathering they form a mixed suite of soils, some fertile and some less so. Metamorphic rocks underlie 17% of the earth's land surface.

Sedimentary rocks underlie 66% of the earth's land surface (Figure 5–3). During earth's history these rocks have been formed from materials eroded from land surfaces and deposited in floodplains, deltas, lakes, and ocean floors. Sedimentary rocks contain material that once was present as soil on the surface of the land and often contain organic carbon forms acquired from plant and animal life that lived on that land. Most but not all sedimentary rocks are relatively enriched with essential elements when compared to crystalline rocks. Six major types of sedimentary rocks can be identified.

Coal is formed from the burial of organic plant and animal residues that accumulated as peat and muck soils (Histosols; see the appendix). Organic sediments that have been subjected to some compaction are known as bituminous, or soft coal; those subjected to more pressure are known as anthracite, or hard coal. Extreme pressures have created oil and natural gas from organic-rich sediments buried deeply in the sedimentary column. The various coals, oils, and natural gases are mineralized forms of carbon

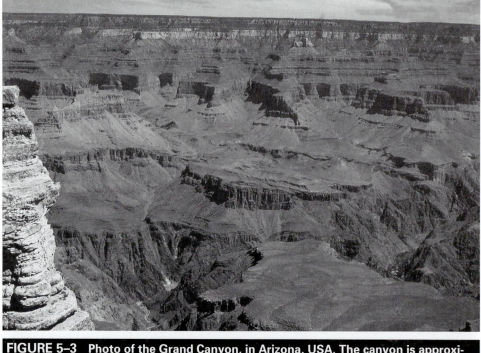

FIGURE 5–3 Photo of the Grand Canyon, in Arizona, USA. The canyon is approximately 1 mile (5,250 feet or 1,924 meters) deep at this point. At the base of the canyon the Colorado River is eroding granite rock. The canyon walls expose layers of sedimentary sandstone, shale, and limestone rocks.

extracted from the air via plants and buried in the geologic column of the earth, and all have their origin in plant and animal life that was present in ancient seas and land surfaces. In some areas recent unburied organic accumulations of peat and muck soils (Histosols) are dug, dried, and used as fuel (Figure 5–4).

Limestone is a carbonate-rich sedimentary rock formed from the residue of plants and animals that once lived in the shallow seas of the world. The major types of limestone are calcite ($CaCO_3$) and dolomite $Ca:Mg(CO_3)_2$. The composition of limestone differs greatly depending on the amount of siliceous material that was buried with the organisms. Most soils formed from limestone are rich in calcium and almost always magnesium. Phosphorus content is almost always relatively high in soils formed from limestone because a major component of limestone is the bone residue of sea animals. Marble is a metamorphosed form of limestone.

Sandstones are hardened sedimentary deposits of coarse-textured sandy sediments that have been eroded from land surfaces and deposited on the bottom of lakes and oceans. Soils formed from sandstones tend to be sandy in texture and thus have low available water-holding capacity. Most soils formed from sandstone are relatively poor in content of essential elements; however, some sandstone contains substantial amounts of lime and greater quantities of essential elements than sandstone that is free of carbonate.

FIGURE 5–4   **Photo of a commercial peat mine in Ireland. The brick-like peat material in the foreground has been compressed and cut by machines and is drying before being gathered for fuel. In the background, sheep are grazing in a pasture formed after approximately 20 feet (6 meters) of peat has been removed.**

Shale is a relatively soft sedimentary rock composed mostly of clay and silt-sized particles. A more compressed and thus harder form of shale is known as *slate*. Like sandstone, the composition of shale and slate is highly variable depending on the composition of the material eroded in the watershed that provided the runoff into the water body in which the material was deposited.

Sedimentary deposits of coal, limestone, sandstone, or shale are now present on the surface of the land where tectonic events have thrust these former sea bottom deposits above the water. Often the layers of sediments that were deposited in level beds are thrust into nearly vertical positions many thousands of feet high, as in some of the major mountains of the world like the Alps, Andes, and Himalayas (Figure 5–5).

Sediments on the land surface that have been deposited in recent geologic time as a result of erosion and deposition are collectively known as *recent sediments*. These areas are primarily located in river floodplains, deltas, and coastal plains. Some recent sediment is composed of unconsolidated surface deposits of clay, silt, and sand, often with considerable quantities of organic residue derived from the erosion of topsoil within the watershed, and they are usually quite rich in essential elements. Some of these areas may be sporadically flooded with water that occasionally destroys the vegetation for a short period of time and may threaten humans. Other geologically recent sediments are at

FIGURE 5–5 Photo of a slab of limestone about 12,000 feet (3,648 meters) above sea level in the Andean mountains of Peru, SA. Note the ripple marks that formed when the limestone was formed under shallow water.

higher elevations and presently not subject to flooding. They are often quite infertile because the minerals that contained essential elements have been decomposed while in a previous soil that was eroded or during transport to the present location.

Finally there is one category of sedimentary materials on the land surface that because of their geologic history of deposition and composition are unique materials for soil formation. These are the *glacial deposits,* especially those in the northern temperate latitudes of Europe, Asia, and North America. Glacial deposits are composed of material moved by continental ice sheets 10,000 to perhaps 300,000 years ago during the ice age (Figure 5–6). These materials are unique among the various geologic materials from which soils have formed in that most minerals contained were never subjected to weathering at the land surface prior to formation of the present soil. The mineral grains were also subjected to physical crushing and grinding by the ice that created fresh mineral surfaces that weather rapidly once exposed to soil formation on the land surface.

All glacial deposits do not contain the same minerals. Glacial deposits that consist of limestone material scraped up by and crushed by the glaciers are chemically rich in most essential elements. This is the case in much of the Midwest of the United States, Eastern Europe, and central China. Glacial deposits consisting of crystalline rock material, as in the northeastern United States, have lesser contents of most essential elements.

The physical properties of glacially deposited materials also contrast. Material directly deposited by the ice in hills known as moraines is known as *till* and consists of a

FIGURE 5–6   **Aerial view of land in eastern Canada. Note the striations of hills, dark color, and valleys, white with snow cover, left as the Pleistocene glaciers passed over the area.**

mixture of particle sizes ranging from rocks and boulders to clay. The continental glaciers were in temperate latitudes, and as climatic conditions warmed during the end of the ice age, summer temperatures caused rapid melting and fast-flowing rivers deposited stratified sand and gravel in broad level areas known as *outwash* plains or deposits. Seasonal freezing conditions in the winter and major movements of the glacial ice created lakes where clay and silt-textured material known as *lacustrine sediments* were deposited.

The outwash plains ceased to be flooded in the winter but were devoid of vegetation. The wind eroded silt-sized particles from these barren floodplains and deposited the silt as a material known as *loess* over the surrounding areas. Thickness of the loess blanket over the underlying material varies from a few inches (Figure 5–7) to tens of feet. Most loess deposits, especially those formed in areas where the glaciers crushed material rich in limestone, have excellent chemical properties and available water-holding qualities associated with silt-sized particles as discussed in Chapter 2. Some of the most productive soils in the Midwest of the United States, Europe, and Central Asia have formed in thick deposits of carbonate-rich loess.

Elements escape the mineral pool only as minerals decompose by processes called *weathering* and enter the available pool. Essential elements move out of, or *weather* from, the mineral pool slowly and only with great expenditures of energy. Release of nutrient elements from the mineral pool requires dissolution of the silicate, carbonate, or oxide minerals that contain an essential element as part of its structure.

FIGURE 5–7 Photo of a soil formed in wind-deposited loess over till deposited by glaciers during the most recent ice age in southeastern Wisconsin, USA. (Arrow indicates boundary between loess and till at 22 inches (56 cm) below the soil surface.)

Because most minerals formed during slow cooling of magma in the earth's crust, or as precipitates at the bottom the sea, they are not stable in soil conditions near the land surface. All minerals are at least slightly soluble in water, some more than others. Small particles weather more rapidly than large particles. Easily weathered minerals are usually more abundant in materials that have never been exposed to conditions near the land surface than in materials that have had previous exposure. The natural fertility of most glacial deposits is in part related to the fact that the minerals were not weathered by prior exposure to conditions at the land surface and crushed by the ice. The silt-sized particles of loess deposits are especially subject to rapid weathering.

As certain minerals weather, they acquire a structural configuration or form secondary minerals that will accept more of certain essential elements than were present in the original mineral. When this happens, available forms of some essential elements are trapped, or fixed (*fixation* in Figure 5–1), in very unavailable forms and return to the mineral pool. One example of fixation is iron-rich silicate minerals that weather to form iron oxides. The iron oxides have a greater capacity to combine with phosphorus than iron silicates and actively compete with roots and microbes for any phosphorus in the available pool. This reaction decreases the amount of phosphorus in the available pool. Fixation is of concern when management attempts to fertilize a soil with an available chemical form of an essential element because some of the fertilizer combines with mineral compounds in the soil and becomes relatively unavailable to plants.

AVAILABLE POOL

The available pool of essential elements exists in the water contained in the soil pores. It is from this pool that plants and microbes in the soil obtain the essential elements they need. Within the available pool both negatively ($^-$) and positively ($^+$) charge ions of essential elements are present. The positively charge ions actively bounce back and forth from negatively charged surfaces of clay and organic particles into water present in the soil pores (Figure 4–3).

As illustrated in Figure 5–1, essential elements enter and exit the available pool via the following six pathways:

1. Enter by weathering from the mineral pool.
2. Enter by the decomposition of organic compounds from the organic pool.
3. Exit by being incorporated into plant roots.
4. Exit by being incorporated into the cells of microbes in the soil.
5. Exit by fixation into the mineral pool.
6. Exit by being leached downward in percolating water.

Not pictured in Figure 5–1 is the human introduction of essential elements into the available pool via soluble inorganic forms of fertilizer.

The movement of essential elements from the available pool into organic life-forms, both through the roots of plants and microbial life in the soil, is of greatest concern to understanding the interaction of soil, land, and life. The movement of essential elements from inorganic ions in the soil to form organic chemicals in plants and microbial cells is fundamental to all life.

Plants and microbes in the soil extract the elements they need while the elements are present in the soil solution. Fueled by the sun's radiation, plant roots search the soil and compete with the microbes in the soil for essential elements available in the soil solution. After ions of essential elements are taken into the plant root, they are transported to the plant cells via the phloem tissue. Within the green cells of plants and in the presence of light energy, carbon dioxide from the air and water combine to form sugar in the process known as *photosynthesis*. The chemical formula for photosynthesis is Light energy + $6CO_2$ + $6H_2O$ → $C_6H_{12}O_6$ (sugar) + $6O_2$. Literally stated, in sunlight the plant uses radiant energy to form sugar from carbon dioxide and water, with oxygen given off as a gas (O_2) to the air. The sugar produced is then transported to other cells in the plant where it transforms and combines with the other essential nutrient elements to create carbohydrates, starches, proteins, and other organic compounds used to grow new tissue such as flowers and seeds in the plant. The various biochemical pathways of individual elements in the growth of plants and formation of the various organic compounds are subjects of numerous volumes of biochemical and physiology research too detailed for this discussion.

Microbes within the soil also remove essential elements from the available pool and incorporate them into their cells. The end result of plant and microbial growth is the assimilation of essential elements as inorganic ions from the available pool into organic compounds of C, H, and O within living cells. Upon the death of the plant and microbial cells, all of the essential elements are part of carbon compounds in organic residue represented by the <A-C> symbols in Figure 5–1. As these carbon residues partially decompose the organically bound essential elements contained enter the organic pool in the soil.

All the ions in the available pool are subject to leaching as water percolates through the soil. The negatively ($^-$) charged and neutral (H_3BO_3) ions are much more susceptible to leaching than are the positively ($^+$) charged ions that are attracted by the negative charges of clay and organic compounds. The potential leaching of nitrate (NO_3^-) nitrogen is of greatest concern. The negatively charged phosphorus ions seldom leach in mineral soils containing iron oxides that fix P ions. It is important to remember that leaching takes place only when water moves downward to a depth where it cannot be reached by plant roots. As you recall from Chapters 2 and 3, leaching is seasonal or sporadic and only occurs when infiltration from rainfall events exceeds the uptake and transpiration of water by the plants.

ORGANIC POOL

The organic pool, as pictured in Figure 5–1, consists of organic compounds in the soil. It is a subtle point, but dead plant remains recognizable as leaves or stems on the soil surface, and root structures in the soil are not analyzed as part of the organic pool and are usually referred to *organic residue*. The <A-C> symbols indicate carbon compounds as organic residue. The aboveground plant parts are most often deposited on the soil surface when the plants or plant parts die. Organic compounds formed in plant roots never leave the soil. Essential elements enter the organic pool as microbes both above the soil and in the soil decompose these organic residues. Microbial tissue is so small that it immediately enters the organic pool as soon as each microbe dies.

As organic residues decompose, the elements that are concentrated in the cytoplasm of biological cells, especially K^+, are easily dissolved in water and return directly to the available pool. The rapid release of K^+ from residue such as leaves on the soil surface is unique. Most other essential elements remain in organic residue until microbes more completely decompose the cells and the essential elements become part of the organic pool as constituents of microbial cells.

As organic residues decompose, their physical shape and form are lost. The resulting organic compounds are large complex molecules that in addition to C, H, and O also contain variable amounts of essential elements. These black- to brown-colored organic compounds with no visible physical structure to identify the original residue result from several stages of decomposition of plant, animal, and microbial tissue and is known as *humus* in the soil.

The activity of billions of microorganisms in the soil is critical to understanding the organic pool. When organic material, either plant residue or manure, is deposited on the soil, all the essential elements contained, except for potassium, are present as organic compounds. Potassium is present in the ionic K^+ form in most residues. The other essential elements as components of humus organic compounds are essentially not available to living plants and microbes. Some plants are known to accept minute quantities of essential elements as very small organic molecules, but this is not a major pathway of essential elements to enter plant roots. A small portion of the organic compounds may be soluble in water and leach if water moves through the soil.

To convert the essential nutrient elements from the organic pool into inorganic ions in the available pool, the carbon structures of the organic compounds must be dismantled. Essential nutrients are released from organic compounds when microbes, bacteria, fungi, protozoa, and actinomycetes attack the carbon structures and combine the carbon and hydrogen in the organic material with O_2 to form carbon dioxide (CO_2) and water. This process is called *respiration*. The general formula for respiration is $C_6H_{12}O_6$ (sugar) $+ 6O_2 \rightarrow 6CO_2 + 6H_2O$ + Energy. Respiration is essentially the opposite of *photosynthesis*. Simply stated, in the process of respiration oxygen combines with sugar and other organic compounds to form CO_2 and water and in the process energy is liberated.

Respiration reactions in the soil are much the same as the fire that we see burn wood and coal at high temperatures, but they take place at low temperatures. Microbes use the energy derived from respiration to fuel their physiological functions in much the same way as humans burn wood, coal, or oil to heat homes or power combustion engines.

The decomposition of organic compounds by microbes in the soil is not completed in one reaction. During each stage of microbial decomposition only about 75% of the carbon in the organic compounds is converted to CO_2 and escapes the soil into the air. Approximately 25% of the carbon is incorporated into the cell structure of the microbes. The life expectancy of individual soil microbes is short, usually a few hours or days. Upon the death of each individual microbial cell, other microbes decompose the dead microbial cells. Each time a generation of microbes die and decompose, essential elements are released into the available pool as inorganic ions, but the new microbial cells need these elements and quickly return them to the organic pool as they incorporate a portion of the released inorganic ions into their cells. Billions of microbes are present in all soils, and because of their profusion throughout the soil they are more efficient in obtaining essential inorganic ions from the available pool than the plant roots. Depending on the composition of the organic material available for microbial decomposition in the soil, microbial activity may hinder or help plant growth. If the ratio of an essential element to carbon in the organic material being decomposed is lower than required in the microbial cell, the microbes will actively obtain that element from the available pool and decrease the amount available for plant roots (see Chapter 4).

Microbial activity, that is, respiration rate, is extremely variable in soils. Soil microbes respire very slowly when it is cold. At temperatures colder than 41°F (5°C), often called biological zero, they essentially cease to respire. There is a multitude of microbe species in soil. In general soil microbes are most active at temperatures of about 95°F (35°C), but individual species each have somewhat different optimum temperature requirements. At temperatures between 41°F (5°C) and 95°F (35°C), the rate of microbial respiration approximately doubles for each 18°F (10°C) rise in temperature. At temperatures above 95°F (35°C), respiration of most soil microbes slows. Most microbes that decompose organic compounds require O_2 from air in the soil. When a soil is saturated with water, their rate of respiration rate decreases. Microbial activity is also influenced by the pH value of the soil with a pH value near 7 being most desirable.

In the total scheme of soil functions, four aspects of nutrient elements in the organic pool are critical:

1. While in the organic pool, almost none of the essential elements are available for direct use by higher plants.
2. Some carbon in the organic matter must be converted to CO_2 and returned to the air before the nutrient elements in the organic pool are converted to inorganic ions in the available pool.

3. Temperature, moisture, pH, and O_2 content in the soil control the rate of the microbial activity that decomposes organic compounds in the soil, making the rate of nutrient release from soil organic matter highly variable and dependent on weather conditions.
4. Microbes are essential to release essential elements from the organic pool but actively compete with plant roots for essential elements in the available pool.

OVERVIEW OF ESSENTIAL ELEMENTS PASSING THROUGH THE SOIL-PLANT SYSTEM

Following the route of elements as pictured in Figure 5–1 is much like driving the highway system on a small island. We always remain on the island unless we get on a boat or an airplane. There are only a limited number of ways essential elements can leave the soil-plant system.

In Figure 5–1 the major escape of essential elements is depicted as the "city" and represents what happens when human food is harvested from the land and transported to concentrations of humans for consumption. The harvest of human food crops accounts for the greatest amount of essential element removal from cropland and is discussed in greater detail in later chapters.

Leaching is the movement of an essential element beyond the deepest extension of plant roots. Leaching is a sporadic event related to rainfall and rate of moisture use by plants. In temperate regions, water for leaching is most available during winter and spring seasons when the soil is cold and plant growth minimal, but it can occur anytime that rainfall fills all the soil pores larger than 0.01 mm in the rooting volume of a soil (Figure 2–6). Anytime an element is in a water-soluble form as an inorganic ion in the available pool or as a water-soluble compound in the organic pool, it can move with that water. Elements that have a positive ($^+$) charge become loosely attached to the negative charge of the clay and organic matter in the soil and are less subject to leaching than those with negative charges ($^-$). Elements like phosphorus adhere to iron and aluminum oxide surfaces that are more abundant in subsoil than in surface soil, and phosphorus leaching is minimal in most soils.

The nitrate (NO_3^-) form of nitrogen has a significant potential to leach. Most often leached elements enter the regional groundwater and ultimately concentrate as a salt or minerals in the ocean. Often elements leached from one soil travel through the groundwater and enter another soil at a different location, a topic for a later chapter. Too much nutrient leaching below the root zone can cause pollution of groundwater and surface water. Some leaching of essential elements is of critical importance in providing essential elements to aquatic organisms. Although only forms of life on the land are considered in this book, we must remember that aquatic life also requires essential elements that are available only from the dissolution of minerals.

There are two other routes for essential elements to leave the soil-plant system not portrayed in Figure 5–1. Nitrogen can escape to the air in saturated soils via the process of denitrification, as discussed in Chapter 4. Erosion of soil material from the soil surface is also a process whereby essential elements are removed from a soil, discussed in later chapters.

To follow the possible routes essential elements may take through the food chain as depicted in Figure 5–1, we must recognize the necessity of *light* for photosynthesis. We must also recognize that *water* is essential to all of the reactions depicted and too much or too little water greatly affects the rate of the various reactions.

Because the greatest number of essential elements, nitrogen excluded, originates in soil minerals, we start our trip around the soil-plant island, as it were, in the mineral pool. From the mineral pool, essential elements dissolve into the available pool. If there is too much water, some of the elements, especially those with negative charges, may leach below the rooting depth of plants and be removed from the system. During dry periods some elements in the available pool may return to the mineral pool via *fixation*.

Both plant roots and microbes in the soil capture essential elements from the available pool. The essential elements captured by the microbes remain in the soil as organic compounds in the organic pool to be again released into the available pool as successive generations of microbes decompose the organic compounds. Some of the essential elements captured by plants also remain in the soil as components of plant roots and enter the organic pool when the roots die and are decomposed. Some of the essential elements captured by plants are transported to the aboveground portions of the plants, combined with carbon the plant has captured as CO_2 from the air and become components of stems, leaves, and seeds. If the plant is allowed to die on site, they are deposited as plant residue on the soil surface.

In their quest for food, humans intercept essential elements by harvesting and transporting plant material for consumption at a location (city) other than the site of plant growth. Humans are often selective and harvest only the plant seed because it contains the most nitrogen and phosphorus and leave only plant parts with lesser contents of essential elements. Animals and birds also tend to transport and concentrate nutrients in nesting sites and resting areas where they preferentially deposit urine and manure.

From manure or plant residue (the <A-C> cells of Figure 5–1), essential elements reenter the soil via three routes:

1. Potassium is easily dissolved from organic residue and quickly goes directly into the available pool as potassium ions (K^+).
2. Several stages of decomposition may take place above the soil until the organic residues are reduced to small particles that are often moved into the soil by insects and worms.
3. If the residue or manure is physically mixed into the soil, microbes decompose the organic tissue in several stages. At each stage of organic residue decomposition, some carbon is vented to the air as carbon dioxide and essential elements are released into the available pool as inorganic ions.

All microbes in the soil obtain energy from the oxidation of organic carbon. As life-forms they also must have the other essential elements, and they vigorously compete with plant roots for the available nutrients in the available pool. The life expectancy of each microbial cell is but hours or weeks, and as each cell dies the essential elements it contains enters the organic pool.

Numerous species of microbes are in the soil, but one group, the *N-fixing microbes,* have a very essential role in supplying nitrogen to the soil by incorporating N_2 from the air into their cells. Although some nitrogen enters the system with rainfall, N-fixing microbes play a major role of incorporating nitrogen into the soil-land-life system. N-fixing microbes also require essential elements from the available pool and fail to fix N_2 from the air if other essential elements are not available. The nitrogen they capture from air is first added to the organic pool and then moves to the available pool as other microbes decompose dead N-fixing microbes. Whether aboveground or in the soil, the central role of most microbes is in the decomposition of organic residue that would otherwise accumulate in and on the soil. To obtain energy for their physiological functions, microbes burn carbon from the organic pool by combining it with oxygen. In this burning process, carbon dioxide is produced that finds its way to the air above the soil. During each cycle of decomposition, approximately 75% of the organic carbon consumed from the organic pool is converted to CO_2 and 25% combines with nutrient elements from the available pool and becomes part of new microbial cells. When individual microbes die, their cells become part of the organic pool.

AMOUNTS AND MAGNITUDES

Having laid out a road map for the elements within the soil-plant system, we turn our attention to a more quantitative look at the amounts of life-essential elements present and how rapidly they move within the system.

The quantity of each essential element that is derived from soil minerals is finite in each soil. The total quantity of nitrogen because of its source in the air can be considered infinite or equally available to all soils. Of the essential elements that must be secured from soil minerals, primarily phosphorus and potassium are discussed because they are most often found to be limiting to plant growth. But only finite quantities of all the other essential elements obtained from mineral sources are present in each soil.

Soils differ greatly in the amount of each essential element present. The total amount of each element present represents the maximum amount of that element that could be supplied to growing plants. Table 5–1 contains total and available amounts of

TABLE 5–1 **Approximate Total and Available Pool (Soil Test) Contents of P and K in 7-inch (18-cm) Layers of Topsoil and Subsoil of Different Soils***

Location (Soil Order)	Depth (inches)	Total P (Lb Ac⁻¹)	Available P (Lb Ac⁻¹)	Total K (Lb Ac⁻¹)	Available K (Lb Ac⁻¹)
West Africa	0–7	625	15	1,518	422
(Alfisol)	16–23	286	3	1,607	211
West Africa	0–7	179	25	625	286
(Entisol)	16–23	107	3	447	70
West Africa	0–7	482	9	3,235	140
(Mollisol)	16–23	179	2	2,411	70
Brazil	0–7	220	3	2,666	70
(Oxisol)	16–23	180	2	2,100	20
USA	0–7	1,729	25	34,095	309
(Mollisol)	16–23	1,493	15	33,057	230

*Amount per acre 7-inch (18-cm) layer of soil is a common unit for agronomists who have to make recommendations for soil amendments. The 7-inch value comes from the standard depth of mixing by a plow or other tillage equipment. Also, a 7-inch thickness of soil over an acre weighs approximately 2 million pounds (907,184 kg). To convert Lb Ac⁻¹ to kg ha⁻¹, multiply by 1.12.

phosphorus and potassium present in 7-inch (18-cm) thick layers of topsoil and subsoil of five different soils. The first four soils were selected to illustrate the low range of these two major elements; the Mollisol in the United States represents some of the most naturally fertile soils in the world.

Total elemental content includes all chemical forms of P and K in the mineral, available, and organic pools. Total essential element quantities are greater in the surface 0- to 7-inch (0- to 18-cm) depth of most soils because near the soil surface substantially greater quantities are present in the organic pool than in the subsoil. In a comparable 7-inch (18-cm) thickness of the subsoil (16- to 23-inch depths), the total amount of each element is almost entirely in the mineral pool. From the examples selected in Table 5–1 we find quantities of total phosphorus ranging from less than 200 to nearly 2,000 pounds per acre 7-inch layer (224–2,240 kg ha^{-1} 18-cm layer). Total potassium contents range from less than 500 to over 30,000 pounds per acre 7-inch layers (560–33,600 kg ha^{-1} 18-cm layer). These ranges are estimated to encompass perhaps 90% of all the soils in the world but do not represent extreme values.

Although the total quantity of P and K in a soil can be quite accurately determined, the measurement of so-called available quantities is much more problematic. The available values given in Table 5–1 are soil test values. Although soil test values are frequently called available, be aware that they are only an index of availability and not precise measurements of available quantities (see Chapter 4, "Determining the Fertility of a Soil"). Soil test available values in Table 5–1 are of value only in pointing out that only small fractions of the total amount of essential elements like P and K can be considered available, and the total amount present generally has little relationship to the available amount. The percentage of the total amount of an essential element that becomes available to food crop plant varies considerably among different soils and weather conditions during the growing season.

Among the many factors that affect the relationship of a soil test value to actual availability are the rates at which essential elements move to and from the mineral and organic pools to the available pool during the growing season. The rate at which elements enter and leave the available pool depends on water content and temperature in the soil, type and size of mineral particles present, pH of the soil solution, and composition of the organic compounds in the organic pool. Estimating available nitrogen is particularly problematic because it is related both to the chemical composition organic matter and the rate of organic matter decomposition in the soil.

There are also differences in the ability of various plant species to obtain elements from the soil. Some plant species form symbiotic relationships with species of microbes known as mycorrhiza. Mycorrhiza are long fibrous organisms that attach themselves to cells in plant roots and are able to densely explore the soil, gather phosphorus and other essential elements, and transmit them to the host plant. This more dense proliferation of tissue capable of ingesting essential elements greatly increases the ability of those plant species to obtain phosphorus.

Most organic residues that enter the organic pool contain all the elements required for plant growth. However, not all organic residues are equal in content of the various elements. To a large extent this is determined by the way plants distribute the elements among the various plant tissues. An example of the elemental concentration within plants is illustrated by analyses of oat grain and straw (stems) in Table 5–2. Phosphorus and nitrogen are approximately four times as concentrated in the grain as in the stem.

TABLE 5–2   Nutrient Element Content of Oat (*Avena Sativa*) Grain and Straw

Element	Oat Grain	Oat Straw
	% of Plant Dry Weight	% of Plant Dry Weight
N	1.7	0.45
P	0.43	0.12
S	0.28	0.33
K	0.64	1.4
Na	0.02	0.3
Ca	0.22	0.9
Mg	0.12	0.1
Fe	0.005	0.0085
Mn	0.008	0.005
Cu	0.0003	0.00023
B	0.00011	0.0007
Mo	0.00016	0.0001

Source: Mengel and Kirkby, 1978.

Potassium and calcium contents are considerably greater in the straw than in the grain. Although relative concentrations differ among different plants, the concentration of nitrogen and phosphorus as protein in the seed tissues is quite universal among grain plants commonly used for human and animal food. It is the seed tissue (grain) that is usually harvested for human food. Calcium and potassium are concentrated in cell wall structures of the stem and leaves of most plants and often left as residue as grain crops are harvested.

The same species of plants also differs in nutrient content per unit weight depending on the concentration of available essential elements in the soil where the plant was grown. Plants grown on a soil with an abundance of a certain essential element usually contain larger quantities of that element per unit weight than plants grown on soil containing lesser quantities. In Table 5–3, note that as the contents of calcium (Ca^{2+}) and potassium (K^+) determined by soil-testing techniques increase, the percentage of each element in the leaves of yellow poplar trees increases.

Similar relationships have been determined numerous times and in many different plant species. The inescapable fact is that plants grown in less fertile soils return less fertile organic residue to the soil upon death of the plant than plants grown on fertile soil.

TABLE 5–3   Potassium and Calcium Concentration in Yellow Poplar (*Liriodendron tulipifera* L.) Leaves Related to Available (Exchangeable) Concentrations in Soil

K^+ in Soil cmol kg^{-1}	K in Leaf %	Ca^{++} in Soil cmol kg^{-1}	Ca in Leaf %
0.21	1.1	3.0	1.90
0.30	1.2	3.1	1.90
0.35	1.3	4.9	1.95
0.40	1.4	5.0	2.05
0.45	1.6	5.1	2.15

Source: Vimmerstedt and Osmond, 1968.

NATURAL VERSUS CROP PRODUCTION ECOSYSTEMS

Human interest in the soil-land-plant system relates not only to its ability to provide human food but also to ecosystems that are less intensely managed. The pools, chemical forms, and pathways that essential elements take in soil-life systems are similar although not entirely the same in all land. Two parameters differ greatly between natural ecosystems and systems of food crop production, often called *agroecosystems*. One major difference is the volume of soil occupied by the plant roots. The other parameter is the amount of time in which the plants are expected to reach maturity.

The physical volume of soil occupied by plant roots is a parameter of the available pool. Depth of rooting can vary from only a few inches of soil for some plants and to great depths for others. In modern agricultural systems, growing food crops estimating an amount of available element via soil tests of the top 7 inches (18 cm) of soil is quite successful. The same degree of success cannot be claimed for tree species where the rooting depth is much greater and the growing period extends over several years.

The rate at which essential elements are required for the growth of most human food crops is much faster than that required for trees and most natural vegetation. The vast majority of human food crops grow from seed to mature plants in approximately 100 days. The food crop plant must grow an entire root system within the soil, stems, and leaves and produce a mature seed during that period of time. With some exceptions natural ecosystems contain perennial species that may be dormant during cold or dry seasons of the year but are able to expand their growth from established root systems whenever growing conditions are favorable. Some plants may grow for several years before they develop seeds necessary for reproducing new plants.

Our common food crops each have somewhat different requirements with regard to the relative quantities of each essential element. Table 5–4 lists the amounts of several essential elements contained in some common food crops. All values presented are for relatively high levels of crop yield per acre of land. Lower yields per acre would either require lesser amounts of essential elements per acre of land or the required amounts of essential elements for the yield of crop listed are taken from a larger unit of land. With the exception of sugarcane, all the crops listed reach maturity in about 100 days. For a comparison between food crops and forest vegetation, elemental contents of both the wood (stems) and total aboveground tree parts in 22- and 60-year-old loblolly pine plantations are also given in Table 5–4. Clearly the rate at which essential elements are available for successful growth of food crops must be much greater than is necessary for tree growth *and* from a smaller rooting volume of soil.

Failure to understand the amount of time food crops and trees have to accumulate essential nutrients seems to be at the heart of a common misunderstanding about soil fertility. Farmers often declare a soil too infertile for crop growth. Statements from farmers and agricultural scientists, such as the soil is "too poor," "infertile" or "chemically degraded," create the impression that nothing will grow. However, growth of trees and many other nonfood plants that require a much slower rate of nutrient element flow from the available pool is often rather vigorous after a cropped field is abandoned because it is too infertile for food crop production.

Crop	Yield	N*	P	K	Ca	Mg	S	Cu	Mn	Zn
	Ac^{-1}					$Lb\,Ac^{-1*}$				
Corn (grain)	150 Bu[†]	135	23	33	16	20	14	.06	.09	.15
Corn (stover)	4.5 T	100	16	120	28	17	10	.05	1.5	.3
Rice (grain)	80 Bu	50	9	8	3	4	3	.01	.08	.07
Rice (straw)	2.5 T	30	5	58	9	5	—	—	1.6	—
Sorghum (grain)	60 Bu	50	11	13	4	5	5	.01	.04	.04
Sorghum (stover)	3 T	65	9	79	29	18	—	—	—	—
Soybeans (grain)	40 Bu	150	16	46	7	7	4	.04	.05	.04
Soybean (straw)	1.5 T	18	2	20	—	—	—	—	—	—
Cabbage	20 T	130	16	108	20	8	44	.04	.01	.08
Potato (tubers)	400 Bu	80	13	125	3	6	6	.04	.09	.05
Tomato (fruit)	20 T	120	18	133	7	11	14	.07	.13	.16
Peanuts (nuts)	1.25 T	90	5	13	1	3	6	.02	.01	—
Sugarcane	30 T	96	24	224	28	24	24	—	—	—
22-yr pine (stem)	28.9 T	51	4	32	46	12	—	—	—	—
(Total)	37.8 T	161	17	80	159	31	—	—	—	—
60-yr pine (stem)	87 T	155	12	121	277	44	—	—	—	—
(Total)	100 T	229	21	173	426	62	—	—	—	—
17-Yr Jungle (ash)[‡]		60	5	34	67	14	—	.3	6.5	.6

TABLE 5–4 Approximate Elemental Content in some Crops, Trees, and Ash

*Values for all elements are in Lb Ac^{-1} for the crop yield given (Lb $Ac^{-1} \times 1.12 = $ kg ha^{-1}).
†Bushel weight depends on grain (i.e., corn and sorghum = 56 lb bu^{-1}; rice =45 lb bu^{-1}; soybeans and potatoes = 60 lb bu^{-1}: T, tons (2,000 lb) per acre; lb, pounds per acre.
‡Elements in ashes after burning 22-year-old Amazon jungle regrowth.

Note: Where both grain and stover or straw values are presented, the values have to be combined to estimate the amount of each element that must be taken by the total plant during its growth. Note that none of these values include the elements contained in unharvested roots. In most food crops, only the grain is harvested for food.

Sources: Anonymous, 1972; Sanchez, 1976; Tew, Morris, and Wells, 1986.

An appreciation for the relative rates of essential element uptake by trees and crop plants can be obtained by dividing the elemental content of 22- and 60-year-old pine trees by 22 and 60, respectively, to obtain an annual rate of uptake. For example, the total amount of phosphorus in a stand of 22-year-old pine is 17 Lb Ac^{-1} (19 kg ha^{-1}), indicating an average annual rate of P uptake of about 0.8 Lb $Ac^{-1}\,Yr^{-1}$ (0.9 kg $ha^{-1}\,Yr^{-1}$). The stover and grain of a 150 Bu Ac^{-1} (9,408 kg ha^{-1}) corn crop requires approximately 39 Lb Ac^{-1} (44 kg ha^{-1}) of P from the soil in approximately 100 days.

Trees are able to grow with considerably lower amounts of essential nutrients available in any given volume of soil, such as the surface 7 inches (18 cm) so often sampled for soil test evaluation for food crops. Also, trees transfer considerable amounts of nutrients from the available pool to the organic pool each year as they grow and shed their leaves, needles, and some smaller branches as organic residue in litter fall.

In perennial natural ecosystems, organic residues from seasonal litter fall of needles, leaves, and small branches decompose and return essential elements to the available pool. The same ions of essential elements may pass through a tree several times during

TABLE 5–5	Elemental Content in Annual Litter Fall from White Pine (*Pinus strobus* L.) and Mixed Hardwoods	
	White Pine	*Mixed Hardwood*
Element	Lb Ac^{-1} Yr^{-1*}	Lb Ac^{-1} Yr^{-1*}
Nitrogen	23.7	30.2
Phosphorus	3.3	4.5
Potassium	4.9	16.1
Calcium	17.1	39.7
Magnesium	2.4	5.8

*Lb Ac^{-1} × 1.12 = kg ha^{-1}.

Source: Cromack and Monk, 1975.

the life of the tree (Table 5–5). The annual or seasonal transfer of essential nutrients by trees and other perennial plants is often referred to as mineral cycling or *biocycling*. Under undisturbed forest vegetation, essential elements obtained from throughout the entire root volume of the trees are concentrated in the surface layers of the soil where the organic residue is deposited. This may enhance the essential element content in the surface layers of soil but reduce essential element content in the subsoil.

Most trees are harvested for wood. Note the relatively small amount of phosphorus in the stems (wood) of pine trees compared to the amount of phosphorus in many of the grain crops. Also note the proportionally higher quantities of calcium in the wood than in the grains. Only the elemental content in the wood (stems) of trees indicates the amount of each element removed by a timber or pulpwood harvest; the total values indicate amounts removed with total tree harvest practices.

ESSENTIAL ELEMENTAL FLOW AND INDIGENOUS SLASH-AND-BURN FARMING

Slash-and-burn subsistence farming is practiced in many tropical jungles today and has been practiced at one time in history in most forested areas of the world. Although the exact timing of farming operations varies to cope with local conditions, all slash-and-burn techniques have evolved to take advantage of nutrient flow dynamics. The scenarios used differ in detail, but in most cases slash-and-burn farmers have adapted the following sequence of operations. Farmers select an area of forest or jungle where there is a substantial amount of aboveground vegetation that will readily burn. They cut and dry that vegetation. They then burn the dried vegetation to rapidly oxidize the carbon and release the essential elements from the organic compounds in the organic residue. They plant their most valuable food crop immediately after the organic residue is burned. This is usually rice or corn depending on the food preferences in that culture. This takes advantage of a relatively large pool of available elements released from insoluble organic carbon compounds as the vegetation burned.

The next crop to be planted, at a time dictated by seasonal conditions of rainfall, is usually a legume, such as beans or cowpeas, which harbor symbiotic N-fixing microbes able to extract N_2 from the air. This compensates for the lower nitrate

(NO_3^-) content now in the available pool because of leaching and removal by the first crop. Also, after a forest is cut leaving the soil unshaded, the daily maximum soil temperatures are higher than while under the shade of the trees (Cunningham, 1963). This accelerates microbial activity and the decomposition rate of soil organic matter. Organic residue of roots and straw from the first crop rapidly decomposed to increase the availability of the other essential elements, ensuring good growth of the legume crop.

After the removal of essential elements in the harvest of the second crop, the content of available nutrients in the soil is low. At that time the slash-and-burn farmer in the warm humid tropics often plants a slower growing crop such as cassava or bananas that may grow for a year or more before harvest. These crops use nutrients slowly released from further organic matter decomposition and the weathering of some minerals in the soil. After the third harvest the field is often abandoned to jungle growth because the release of nutrient elements from the mineral and organic pools is too slow for adequate food crop growth and the site is declared "infertile." Growth of weeds and trees that require a less rapid supply of nutrients than food crop plants is usually rapid in the abandoned area. Sustainable slash-and-burn practices allow natural forest vegetation to grow until the amount of biomass in the natural vegetation contains enough essential elements that the ashes will supply sufficient nutrients for another cycle of slash and burn. In many places this takes from 10 to 30 or more years.

OVERVIEW OF ELEMENTAL TRANSFERS IN SOIL, LAND, AND LIFE

Comparisons of soils on the basis of their total mineral composition or organic matter content often fail to predict plant growth. Attempting to grow a food crop on a soil that has only a large supply of nutrients in the mineral and organic pools presents the same problem faced by a thirsty person standing in front of a coin-operated vending machine with only $100 bills in hand. A plant root, like a vending machine that will not accept $100 bills, will not accept nutrient elements locked in organic compounds or silicate minerals. The essential elements have to be in the available pool. The amount of each essential element that is available to the desired plant depends on many variables of temperature, moisture availability, chemical composition of both the mineral and organic pools, the volume of soil explored by the roots, and the length of time available for the desired plant to grow.

How capable are different soils of supplying the essential elements for plant growth? Total elemental contents do identify a finite maximum limit of natural soil fertility. Comparing the content of essential elements in harvested food crops, Table 5–4, with total contents of those same elements in various soils, Table 5–1, illustrates the finite limit of any soil to produce a food crop. Note that as few as ten high-yielding grain crops would harvest and remove *all* the phosphorus in a 7-inch (18-cm) topsoil layer of many soils. Even the most fertile soils contain only enough phosphorus for about 100 crops. Total elemental removal by cropping is unattainable by any known technology. Total removal of all nutrient elements from any soil by cropping would result in a sterile soil incapable of growing any vegetation to protect it from erosion. Fortunately, the chemical balance among the various pools never allows complete removal of essential nutrient elements so a soil never becomes totally sterile. Although total nutrient removal is unattainable, and undesirable, calculating total nutrient elemental removal by cropping demonstrates that no soil is an infinite provider of essential elements derived from soil minerals, especially phosphorus. Every soil has a finite amount of those essential elements derived from soil minerals.

Although all plants require the same elements, plant species differ in the relative proportions of elements required. Plant species in a vegetative community naturally change with time, even if undisturbed by people. Fire, wind, and disease are agents that affect all plant communities and may drastically alter the plant components of the ecosystem. The belowground ecosystem is altered in response. There appears to be enough checks, balances, and feedback mechanisms that the soil-plant system seldom, if ever, fails to function. A soil destroyed for the support of one plant community continues to function supporting a somewhat different plant community. Vegetative changes on a parcel of land are not synonymous with impending doom of a soil. A soil that is bankrupt for crop growth may sustain a good growth of natural vegetation that has a much slower rate of elemental uptake from the available pool.

LITERATURE CITED

Anonymous. 1972. Remember the Plant Food Content of Your Crops. Better Crops with Plant Food. Vol. LVI/1. pp. 1–2. Potash Institute of North America, Atlanta, Georgia.

Blatt, H., and R. L. Jones. 1975. Proportions of Exposed Igneous, Metamorphic, and Sedimentary Rocks. *Geological Society of America Bulletin* 86:1085–1088.

Buol, S. W. 1995. Sustainability of Soil Use. *Annual Review of Ecology and Systematics* 26:25–44.

Cromack, K., Jr., and C. D. Monk. 1975. Litter Production, Decomposition, and Nutrient Cycling in a Mixed Hardwood Watershed and a White Pine Watershed. In F. G. Howell, J. B. Gentry, and M. H. Smith (Eds.), *Mineral Cycling in Southeastern Ecosystems* (pp. 609–624). Tech. Inf. Center. U.S. Energy Research and Development Administration. U.S. Department of Commerce, Springfield, VA.

Cunningham, R. K. 1963. The Effect of Clearing a Tropical Forest Soil. *Journal of Soil Science* 14:334–345.

Mengel, K., and E. A. Kirkby. 1978. *Principles of Plant Nutrition.* Berne, Switzerland: International Potash Institute.

Sanchez, P. A. 1976. *Properties and Management of Soils in the Tropics.* New York: Wiley.

Tew, D. T., L. A. Morris, H. L. Allen, and C. G. Wells. 1986. Estimates of Nutrient Removal Displacement and Loss Resulting from Harvest and Site Preparation of a Pinus-Taeda Plantation in the Piedmont of North Carolina, USA. *Forest Ecology and Management* 15(4):257–268.

Vimmerstedt, J. P., and C. A. Osmond. 1968. Response of Yellow-Poplar to Fertilization. In C. T. Youngberg and C. B. Davey (Eds.), *Tree Growth and Forest Soils* (pp. 127–154). Corvallis: Oregon State University Press.

CHAPTER REVIEW PROBLEMS

1. Why do soils formed from limestone usually contain greater quantities of essential elements than soils formed from igneous rocks?

2. Which two elements combine and escape from the soil when essential elements are converted from the organic pool to the available pool?

3. Which essential elements are more concentrated in grain than in straw (plant stems) of most food grains? How does this affect the fertility value of crop residue "recycling" as a fertilizing technique?

4. What happens to nitrogen in the available pool when organic material with a very high carbon-to-nitrogen ratio is added to the soil?

5. If a soil contains 1,000 pounds of total P per acre and 10 pounds of that P is taken up by a crop during a growing season with half going into the harvested grain and half into the straw left in the field as residue, how many crops can be harvested before all of the P is removed from the field?

6. Using values given in Table 5-4, determine the pounds of P removed during 22 years of growing 150 Bu of corn per acre per year. How much more or less is this than the amount of P removed when a 22-year-old stand of pine is harvested for pulpwood?

CHAPTER

6

Soils and Land with Unique Features

Most humans live in areas where the land and soil have properties compatible with the ability to grow food. Therefore soils and land most familiar to human experience are not extreme in properties. Familiarity with only a few soils has fostered many widely held concepts of sameness among soils. The earth has approximately 33.1 billion acres (13.4 billion ha) of land. Approximately 11% of that land, 3.7 billion acres (1.5 billion ha) is presently cultivated for human food production. It has been estimated that 25% of the land, 8.3 billion acres (3.4 billion ha) could be cultivated if all presently known technology were utilized. Some conditions render soils unusable for food production even with all known technology. Many soils that could be used for food production require management inputs that are well known but are not economically feasible under present conditions in many parts of the world.

Soils and land that fulfilled human needs for crop production with little technologically based management became centers of early civilizations. Today most of these areas are densely populated. Extensive areas of the world require capital investments in infrastructure and technology before they can be successfully cultivated. Such lands are sparsely populated and little studied. In this chapter both lands where soil conditions have historically supported bounteous food production with rigorous cultivation techniques and lands presently little used for food crop production are discussed.

COLD BEER BUT NO POPCORN: PERMAFROST

In the coldest areas of the earth, such as the interiors of Antarctica and Greenland, there is a permanent ice cover and no soil is present. These ice-covered areas are not considered land. Approximately 13% of the land in the world has permanently frozen subsoil. This condition, widely known as permafrost, dominates in polar latitudes of the Northern Hemisphere. Small areas of land with permanently frozen subsoil are also present at very high elevations throughout the world. Soils with permanently frozen subsoil are present where the mean annual air temperature is less than about 16.7°F (−8.5°C). Permafrost is present in less cold areas if the soil surface is covered with plant residue or vegetation such as sphagnum moss that insulates the soil from warming from the sun.

The so-called permanence implied by the name *permafrost* is often misleading. In many cold polar latitudes, long days during the brief summer warm the surface layers of the soil enough for plant growth although the subsoil remains frozen. If the soil surface is covered by organic material, a good insulator, the soil may not thaw or only thaw to a shallow depth during the summer. However, when surface vegetation or organic residue is removed by fire or cutting of the vegetation, summer heat thaws the soil to a much greater depth. Many people have attempted to improve a good trail or a primitive road by clearing the vegetation and organic residue on the soil surface. After clearing they find that what was a firmly frozen roadbed becomes a mire under the heat of the summer sun after the organic residue blanket is compacted or removed. Much of the insulating ability of organic residue is lost when compacted. Fire often destroys the organic residue on the soil surface with a similar result. The permafrost is restored when native vegetation reestablishes itself in the burned or otherwise disturbed area and a layer of organic residue is reestablished on the soil surface. This may take several years or decades given the slow growth of vegetation in the short growing season of polar latitudes.

Even though the soil is permanently frozen at a shallow depth when covered with organic debris, a cleared soil surface may warm enough to permit growing of some food crops during the long daylight hours of the brief summer in some polar locations. In a few locations vegetable crops like radishes and cabbage that can mature in 30 to 60 days can be grown, but grains, such as corn and wheat that require more time to mature, are not possible. Inhabitants rely heavily on fish and migratory animals for food. By digging a cave into the permafrost and then making sure the soil surface above is well insulated from summer heating, a natural household freezer is created. The permafrost provides a natural refrigerator to keep their beer cold, but their popcorn must be imported.

TOO DRY TO TAKE ROOT

Water is necessary for biological activity. Actively growing plants require some available water in the soil material surrounding their roots. The soils on approximately 23% of all land in the world are considered arid. A practical definition of arid land is where in normal or average years there are less than 90 consecutive days of plant-available water from rainfall and available water in the soil during the time temperatures are warm enough for plant growth. Although the 90 consecutive days of reliable plant-available moisture may appear arbitrary, it is a good approximation of the land where cultivation of common food crops is not possible without irrigation. Of course arid soils do support some natural vegetation adapted to water conservation. The types of natural vegetation present differ, often in response to temperature, but collectively they can be referred to as desert vegetation and seldom provide much human food.

Many plants are remarkably capable of protecting themselves for long periods of time without obtaining water, but none can exist when absolutely no water is ever available. One way annual or seasonal plants exist with minimum amounts of water is by quickly producing a seed when water is available. Such seeds are usually well protected from water loss by a thick outer coat that is able to withstand prolonged dryness. The seeds may lay dormant in the soil for a year or more but germinate in the few days after being wetted and produce a new plant that will rapidly grow, flower, and produce seed.

Perennial plants sparsely populate many arid lands. Plants that grow on arid land reduce their consumption of water to a very low level and grow very slowly but survive long periods of soil dryness. Most perennial plants in arid lands develop an extensive shallow root system that extends laterally around the plant so as to intercept as much water as possible from infrequent rains. Once a perennial plant is established, it effectively curtails

FIGURE 6–1 **Photo of vegetation dominated by widely spaced saguaro cacti and creosote bush in the Sonora Desert of southern Arizona, USA.**

the growth of other plants in the area occupied by its root system. The result is a wide spacing of plants and appreciable areas of bare soil between plants (Figure 6–1).

Soils in arid areas have remarkably diverse properties and characteristics. A high proportion of soils in arid areas have abundant contents of essential elements in the mineral pool. However, with a limited supply of water available for plant growth, little organic residue is added and only small amounts of organic carbon and essential elements are present in the organic pool. Natural plant density is low in deserts. This leaves a high proportion of the soil surface unprotected from raindrop impact, assuring rapid soil erosion from upland slopes and deposition of nutrient-rich sediments and run-on water on the floodplains. The dependence of natural vegetation on the accumulation of run-on water is easily seen in the floodplains during the driest parts of the year (Figure 6–2).

Some of the first civilizations developed in warm arid areas of the world. Water was the limiting factor for food production and the growing of food was, and in large part still is, limited to only a small portion of arid land where it is technically feasible and economically possible to irrigate.

In arid lands many floodplains are nearly level, and the soils are usually deep, friable, and free of stones. Although humans often have to evacuate the floodplains during flood events, they can return and plant desired food plants in moist fertile soil. Floodplains are always adjacent to rivers. Early humans were able to dig channels in the friable, rock-free soil of the floodplains and direct the flow of water from nearby rivers to cultivated fields. If water continues to flow in the river after secession of the flood, it

FIGURE 6–2 Photo of natural vegetation in southern Arizona, USA, in midsummer. Note taller green vegetation along the dry riverbed where during infrequent rain events runoff water from adjacent upland provides additional water.

is possible to divert some water away from the river in networks of constructed ditches. This increases the amount of land used for food crops and the length of time water is available to crops. To facilitate the flow of water into a network of constructed ditches, various devices were constructed to lift water from the main ditches into a series of smaller ditches and provide water to cropped fields for periods of time sufficient to grow food crops (Figure 6–3). In some floodplains the water table is relatively near the soil surface and it is possible to dig wells from which to pump water for irrigation. Almost without exception arid regions have few clouds, thus they have abundant solar radiation to power photosynthesis. When provided with irrigation water, very high yields of many food crops are obtained on irrigated land in arid regions of the world.

Rivers adjacent to the floodplains also provide the infrastructure needed to transport food. Where it was possible for early humans to produce a reliable amount of food in addition to their immediate needs, and transport that food via boats, concentrations of people free of the daily labors of food production could be fed. This enabled humans to develop more intricate urban societies.

It is interesting to speculate about the influence seasonal temperature cycles and seasonal return of the floods along the Nile, Tigris, and Euphrates Rivers had in organizing nomadic hunters and gatherers into societies that learned to store food. Did the regimentation of the seasonal flooding condition the people to accept regimentation of government? Did the common need for water that could be provided

FIGURE 6–3   Photo of an oxen-powered Noria lifting water from a canal to irrigate adjacent fields in Egypt.

by the construction of irrigation canals promote cooperation? Did the essential nutrient richness of the soil deposited by the flooding water assure that good physical health was sustainable at one location and enable the expenditure of muscle power required to build great temples and pyramids? Perhaps all of these conditions combined to shape the culture of early civilizations.

TOO MUCH SALT: HIGH BLOOD PRESSURE

There is one inescapable lesson from the history of human attempts to grow food in arid lands with the use of irrigation. That is an often-repeated scenario of salt accumulation in the soil resulting in the demise of agricultural productivity (Hillel, 1991).

Soils with a high content of soluble salts are called *saline* or salt-affected soils. In arid regions there is little or no leaching from natural rainfall, and most of the water stored in the soil after a flood or irrigation event is lost to the atmosphere via evaporation and transpiration. As water evaporates soluble salts contained in the soil water concentrate on or near the soil surface. In more humid areas the concentration of soluble salt in the soil water remains low as rainfall periodically leaches through the soil and the soluble salts are removed from the soil. The leached salt enters the groundwater and eventually reaches the oceans and large inland seas of the world where the salt is concentrated by evaporation, creating salty or brackish water.

Easily dissolved chloride and sulfate salts of calcium ($CaCl_2$; $CaSO_4$), magnesium ($MgCl_2$; $MgSO_4$), and sodium ($NaCl$; Na_2SO_4) are the most common salts that accumulate in salt affected soils. Occasionally nitrate salts may also contribute to the problem. Although small quantities of these salts are present in most soils, they become a problem only when their concentration becomes so high that soil water does not flow easily into the plant roots.

Water naturally flows through the surface of a plant root from soil water with low salt concentration into solutions of higher salt concentration within the plant root by a process known as *osmosis*. As the amount of salt dissolved in the soil water increases and approaches the salt concentration in the cells of the plant root, this process slows. Although some plants are adapted to grow in salty water, essentially no food crops are adapted to grow in salty water such as ocean water. Some food crops are better adapted to higher concentrations of salt than others, but none are able to survive where the salt concentration in the soil water is high.

All water, after contact with soil or geologic rock, contains some soluble salt. The amount of soluble salt that plants can tolerate in soil differs depending on both the type of plant and to some extent the stage of plant growth. Young plants usually are more susceptible to salt damage than mature plants. Salt content in soil is usually determined by measuring the electrical conductivity of a saturated soil sample. In most cases a soil is considered saline if the electrical conductivity exceeds 4 millimhos cm^{-1} (0.4 Siemens per meter) at 25°C. This concentration of salt is approximately equivalent to an osmotic pressure of 0.1 MPa (approximately 1 atm), which is additive to the capillary pressure (tension) created by the soil pores (see Chapter 2).

In arid regions, some saline soils naturally develop in landscape depressions where the water table is near the surface, water is lost via transpiration and evaporation, and soluble salts accumulate. This condition is most common on the nearly level floodplains near rivers in arid climates. The saline conditions of greatest concern to food production are created by improper use of irrigation water. Much and perhaps most of the salinity problems that have plagued irrigated agriculture in arid lands have resulted from adding water to the soil, thereby elevating the water table, and a failure to provide a drainage system for controlling the water table depth and leaching the soluble salts from irrigated land.

A plausible explanation for this historically repeated failure on the part of humans practicing irrigated agriculture is that the benefits of irrigation are quickly realized. The effects of salt accumulation in the soil usually do not become apparent until the addition of irrigation water causes the water table to rise into the rooting depth near the soil surface. As the water table approaches the soil surface, capillary water moves upward, is evaporated and the soluble salts dissolved in the water precipitate on and near the soil surface in the rooting zone of the crop plants. The presence of excessive soluble salt can sometimes be seen as a white crust or effervescence on the soil surface, but often this is visible only after the plants have severely wilted or died from lack of usable water (Figure 6–4).

In most areas it takes several years of irrigation before salts accumulate in sufficient quantity to impair plant growth. During the initial years of irrigation, yields are high. Within an area of irrigated land, saline problems first appear in slight depressions of the land surface where excessive irrigation water has caused the water table to approach the soil surface. Prior to an understanding of the effect of salt on crop growth, these areas of crop failures were probably attributed to other causes, and if additional land was available the salted land was abandoned and irrigation waters diverted to adjacent unsalted land. As long as irrigated crops flourished and new areas were available for irrigation,

FIGURE 6–4   Photo of white effloresces of salt crystals formed on the soil surface as saline water evaporated.

little attention was given to the accumulation of soluble salt in the soil. As human populations increased and irrigation intensified, the water table rise and associated salt accumulation spread to larger and larger areas. As the water table approaches the soil surface from prolonged irrigation, it is necessary to construct drainage systems to lower the water table and provide for the leaching and removal of soluble salt from the affected irrigated area. This is expensive, laborious, and usually requires the cooperation of individuals in adjacent land. In many soils leaching is not easy to do because of the presence of nearly impermeable clay-textured subsoil near the soil surface. In such cases artificial drainage systems are difficult to install, and removal of salt is slowed by the slow permeability of the soil. Even well-designed irrigation systems with adequate and sustainable sources of irrigation water and properly constructed drainage systems to provide for the leaching of salt pose problems for surrounding areas. Salt-rich water discharged from a drainage system, sometimes called *tailing water,* is undesirable for use in irrigation and thus considered a waste product that must be disposed of, usually by discharge into an ocean or a salty inland sea. Many lawsuits, and more severe hostilities, have resulted when the discharge of salty water drained from irrigated fields was allowed to flow into a water supply that farmers downstream were using for irrigation.

All soils can become salty if soluble salts are applied. The common concern for so-called burning of lawn grass or a flower bed with an overapplication of fertilizer is in reality an increase in soluble salt content of the soil water from the soluble salts present in the fertilizer. Potted plants can suffer from salt damage when water is added for long periods of time and the container is not leached to remove the soluble salt. Remember that much of our urban drinking water is chlorine enriched, and chlorine salts are some of the most common components contributing to saline soils. Even if rainwater or distilled water (water prepared by evaporation and condensation processes to remove all soluble salt) is used, some soluble salts will accumulate in a flowerpot from dissolution of plant food (fertilizer) or the minerals contained in the potting soil. Periodic leaching with low salt content water is desirable to guard against the creation of high osmotic pressure both in irrigated land and flowerpots.

Most river water has a low content of soluble salt. In many areas where there is no access to river water, irrigation water is pumped from deep underground aquifers. However, the water in many deep underground aquifers has appreciable quantities of soluble salt dissolved from the geologic minerals through which it passes. In these areas the water table is deep and excess water is annually applied to leach the soluble salts to depths below the rooting zone prior to the planting of crop plants. The leached salts may eventually reach and increase the salt content of the groundwater but do not substantially elevate the water table into the root zone. Many underground aquifers contain a finite quantity of water stored over centuries and are quickly depleted when extensively pumped by irrigation wells. The depletion of groundwater aquifers is a frequent cause for the failure of irrigation systems. A common scenario in irrigated areas where deep groundwater is used for irrigation is that when a relatively shallow aquifer of groundwater is found and successfully tapped for irrigation, people congregate in the area. More wells are constructed to tap the aquifer and more land is irrigated. As the supply of water in the aquifer is pumped more rapidly than the aquifer is recharged, the depth of the wells must be increased. It takes power to pump water from a deep well, and the deeper the well the more costly it is to pump the water. Often farmers facing increasing depth to water in their irrigation wells survive for several years by using minimum amounts of water, neglect the leaching necessary to remove the salt from the root zone, and thereby create saline conditions. As pumping costs increase, farmers can no longer profitably grow crops and the land is abandoned for irrigated crops.

Salinity, the scourge of irrigated agriculture throughout recorded history, can be avoided if appropriate long-range planning and investments are made in the construction and operation of the irrigation systems. A reliable and sustainable source of irrigation water must be assured. There must be enough irrigation water available to provide for the removal of salt via leaching to a drainage system. It is also imperative that the water collected in the drainage system be cleansed or has a corridor to the ocean or other saltwater body without interfering with humans or natural ecosystems along the way (Letey, 1994).

TOO MUCH ALKALI: SOFT SOAP TREATMENT

Alkali (also called Sodic) soils, like saline soils, are almost all found in arid or semiarid areas. Alkali conditions are caused by a high content of sodium ions (Na^+) in the soil. In some cases, magnesium ions (Mg^{2+}) may also contribute to the alkali condition. Alkali soils form primarily where there is a high concentration of rather soluble sodium minerals in the geologic material from which the soil forms. Most sodium salts are soluble. Sodium chloride (NaCl) is common table salt. Sodium is also an element that causes soap to be effective in dispersing the dirt on our dirty clothes and dishes.

BOX 6-1

Schematic Representation of the Effect of Sodium, Calcium, and Aluminum Ions on Clay Particles in Soil

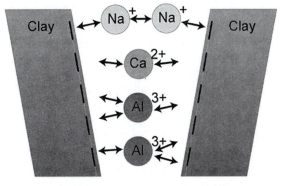

In the schematic only one or two ions of each element are shown between the clay particles, but it is well known that numerous ions are in that position. However, elements like sodium with only one positive charge are less closely attracted to negative charges (CEC) on the clay and organic matter (not shown) particles. Therefore, the clay particles are further apart. When ions like calcium with two positive charges are present, the ions between the clay particles are denser and the clay particles are brought closer together. Ions with three positive charges more strongly attract the clay particles and tend to bring the clay particles even closer together.

When sodium ions constitute more than about 15% of the ions present in a soil, the clay and organic matter particles tend to disperse and are easily suspended in water. Exchangeable sodium percentage is determined by (Na^+ $cmol_c$ kg^{-1}/$CEC_{8.2}$ $cmol_c$ kg^{-1}) $\times$ 100 = Exchangeable sodium percentage. (See $CEC_{8.2}$ in Box 4-2.)

When sodium ions (Na^+) concentrate in soils to the point where they make up more than about 15% of all the basic cations (Mg^{2+}, Ca^{2+}, K^+, Na^+) on the clay and organic matter surfaces, clay and small organic particles become suspended in the soil water (Box 6-1). As the soil dries the small clay and organic particles are deposited on the surface of the larger voids, effectively clogging them and greatly reducing the rate at which water can move in the soil.

In areas where rainwater leaches through the soil during some part of the year, sodium that weathers from the sodium-bearing minerals is normally leached to the groundwater and eventually concentrates in oceans and inland seas. Naturally occurring areas of soils with too much sodium are frequently known as "slick spots" because they remain wet and slippery much of the time as the clay and organic compounds suspended in the sodium-rich water concentrate on the soil surface. The visual appearance of black organic compounds on the soil surface has given rise to the name *black alkali* for such soils in some parts of the world.

Although most soils with excess sodium form in arid or semiarid areas where the geologic material is rich in sodium-bearing minerals, the disposal of sodium-rich wastewater has created alkali soil conditions when discharged onto land in humid areas. High sodium content in wastewater can result from various industrial processes and cleaning operations where soap and other sodium-rich materials are used. The presence of excess sodium usually becomes apparent only after the soil becomes less and less permeable to the infiltration of water. With time the surface soil becomes sandy as the clay particles are dispersed and carried downward to form a dense clayey subsoil layer that is nearly impermeable to water movement.

The presence of excessive amounts of sodium can often be detected by high pH values in the soil. Soil pH values above 8.5 are a clear indication that excess sodium is present. High pH values also decrease the availability of several essential nutrients within the soil (Figure 4–4). The most affected nutrients are iron, manganese, copper, and zinc. Most plants need very small amounts of these elements, but it may become necessary to apply them as foliar sprays when soil pH values are high.

It is difficult to correct the physical clogging of the soil voids in a soil that has resulted from too much sodium. Adding gypsum ($CaSO_4$) to irrigation water and leaching the soil so calcium can replace sodium on the clay surfaces can attain partial reclamation. This may take several years because the subsoil, clogged by clay, may have very slow permeability.

THAT BITTER TASTE: ALUMINUM TOXICITY

Frequently referred to as simply "acid soil," many soils in the world have concentrations of aluminum ions (Al^{3+}; $Al(OH)^{2+}$; $Al(OH)_2^+$) in the soil solution (Figure 4–3) of the available pool that are too high for the growth of many desirable food crop plants. This condition is always associated with an abundance of hydrogen ions (H^+) and thus detectable by acid soil pH values. Aluminum and hydrogen ions displace calcium, potassium, and magnesium on the negatively charges surfaces of the clay and organic particles. Note the decrease in availability of these elements as pH values decrease in Figure 4–4.

It is not possible to specify an exact concentration of aluminum ions that is toxic to plant roots. Some crop plants are more tolerant than others, and some plants like tea and rubber thrive in soils with high aluminum concentrations. Many aluminum-tolerant nonfood crop species grow well in acid soils. Also, among some of the most widely grown food crops, several advances in plant breeding have created cultivars or hybrids that are more tolerant of acid aluminum-rich conditions in the soil. Many of the common food plants are comfortable when the aluminum ions constitute less than 60% of the exchangeable ions held on the clays and organic matter in the soil (Box 6-2). Soils with pH values greater than 5.2 probably contain less than 60% aluminum as exchangeable ions in the available pool. At soil pH values less than 5.0 to 5.2 in mineral soils, aluminum-toxic (acidic) conditions can be expected to reduce the growth of many crop plants. Notable crop plants that are the least tolerant of aluminum concentrations are alfalfa, soybeans, and peanuts. Most cultivars of these plants are adversely affected when aluminum ion concentration exceeds 20% of the ions held on the exchange sites of the soil clay and organic matter. This level of aluminum is approximated by soil pH values less than about 5.6.

BOX 6-2

Aluminum-Toxic Calculations

To determine the level of extractable aluminum in a soil that may be toxic to plants, the following procedure is most often used.

The exchangeable Ca^{2+}, Mg^{2+}, K^+, and Na^+ ions in the sample are displaced by an unbuffered ammonium chloride (NH_4Cl) solution, and the Al^{3+} ions are displaced by 1 N KCl solution (ECEC in Box 4-2). The quantity of each element is determined by chemical analysis and expressed as $cmol_c$ kg^{-1} of soil or mEq $100g^{-1}$ of soil. From these values, the percentage of aluminum is determined by the following formula:

$$\% \text{ Aluminum saturation} = [Al^{3+} \, cmol_c \, kg^{-1} / (Ca^{2+} + Mg^{2+} + K^+ + Na^+ + Al^{3+}) \, cmol_c \, kg^{-1}] \times 100$$

Aluminum-toxic conditions are most often present in soils with no carbonates in the mineral pool. Most acid soils are formed from acid igneous rocks, acid shale, or sediments derived from such materials that contain few if any carbonates and are present mainly in humid areas where precipitation exceeds evapotranspiration and leaching occurs during some part of most years. In most of these soils many of the essential nutrient elements present in the geologic material have been partially removed by weathering below the soil and/or in the geologic transportation processes that have moved the material over the millennia. Many acid soils were naturally forested, and centuries of plant residue have enriched the surface soil with essential elements relative to the subsoil. Aluminum-toxic conditions do form in surface soils but are frequently more severe in the subsoil.

In acid soil, aluminum ions (Al^{3+}, $Al(OH)^{2+}$, or $Al(OH)^+$) are concentrated in the available pool and enter the roots of susceptible plants. There they cause cell walls to enlarge, slowing the flow of cytoplasm. The aluminum-affected roots do not grow normally and usually develop a knotty shape (Figure 6–5).

In strongly acid, severely aluminum-toxic subsoil, the content of nutrient elements in the available pool, especially calcium ions, is very low. The growing root tips of most plants cannot enter soil material that does not contain some available calcium ions (Ca^{2+}). Plants are not able to translocate calcium to growing root tips even if calcium, obtained from surface soil, is present in the plant. Often the aluminum-toxic condition reduces or precludes, depending on its severity and the sensitivity of plant in question, the elongation of roots into the subsoil. This restricts the root system of aluminum-sensitive crop plants to a shallow depth and reduces the amount of water available during periods of drought. Shallowly rooted plants are more susceptible to drought than more deeply rooted plants. In some soils extremely acid conditions extend to the surface soil, and many types of desirable crop plants cannot be successfully grown until the acid condition is corrected by additions of chemically basic material.

The most common material used to correct acid soil conditions is finely ground limestone, either calcite ($CaCO_3$) or preferably dolomite ($Ca:Mg(CO_3)_2$). This management practice is commonly called *liming*. It increases the pH value in the soil and thereby increases the availability of many of the essential elements in acid soils

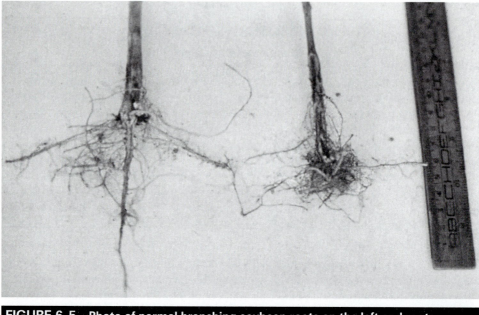

FIGURE 6–5 Photo of normal branching soybean roots on the left and roots affected by toxic concentrations of aluminum on the right.

(Figure 4–4). Although liming is a common practice in modern agriculture, the same effect can be achieved to some extent by burning large amounts of biomass as is practiced by slash-and-burn farmers in tropical jungles where there are extensive areas of aluminum-toxic soils.

Unfortunately, the pH-altering effect of lime or the neutralizing effect of ash does not persist for more than a few years. Lime application must be repeated because the soil continues to acidify as growing plants exchange hydrogen ions for the nutrients they need in the available pool (Figure 4–3) and hydrogen ions are released from decomposing organic matter. Liming materials are very slow to move from the soil surface into the subsoil. To immediately create a deep root zone in an acid soil, the liming material must be physically incorporated into the soil by tillage operations. Evidence does exist that with continued applications of lime over several tens of years calcium does move downward and reduce the aluminum toxicity in the subsoil in naturally acid soils (Figure 6–6).

Aluminum-toxic conditions can also be created by the use of acid-forming fertilizers in the plow layer of soils that have little or no carbonate, but this condition is easily corrected by physically incorporating ground limestone when the cropland is plowed. In most areas where fertilizer is available, farmers are able to obtain lime and apply sufficient quantities every few years to compensate for the acidifying effect of the fertilizer. However this is not universally true. Although it may require only a few hundred pounds of fertilizer to adequately fertilize an acre of cropland, the quantity of lime required is almost always more than 1 ton per acre (2,240 kg ha^{-1}) and may be much greater depending on the amount of exchangeable aluminum present. Although the cost of crushed limestone is much less than fertilizer, the physical transport and

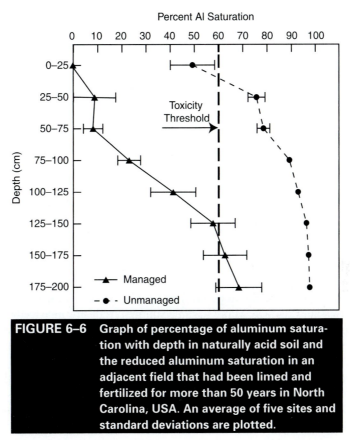

FIGURE 6–6 Graph of percentage of aluminum saturation with depth in naturally acid soil and the reduced aluminum saturation in an adjacent field that had been limed and fertilized for more than 50 years in North Carolina, USA. An average of five sites and standard deviations are plotted.

Source: Buol and Stokes, 1997.

application of such large quantities can be prohibitive. In some parts of the world, finely ground limestone is not available either because of an absence of limestone in the area or failure of the people to develop an industry for crushing limestone. In many farmers' fields and even in controlled experiments, the benefits of chemical fertilizers are sometimes not realized because of a failure to control soil acidity with applications of lime, and the crops fail to respond to the added fertilizer. In all soils not rich in carbonate minerals, continued applications of lime are needed to neutralize both natural acidity buildup and acidity introduced with most chemical fertilizers.

Aluminum-toxic conditions are not visible to the human eye. Locally they may sometimes be identified by the presence of certain indicator species in the natural vegetation. A chemical analysis of the soil is required for verification. Keeping a good record of soil pH values from year to year via soil test technology and making sure the soil pH values are maintained above 5.2, or above 6.0 if acid-sensitive crops are to be grown, is a requirement for sustainable cultivation of soils that do not have carbonate minerals present in the mineral pool. Care must be taken not to apply too much lime and increase the pH value much above 7.0 and thereby decrease the availability of iron, manganese, boron, and other essential elements (Figure 4–4).

EMBALMED PLANT REMAINS: PEAT AND MUCK

When natural ecosystems are undisturbed, the life cycles of the plants and animals living on the soil are short compared to the longevity of the soil. The carbon that plants ingested from the air is deposited as a major component of the organic residue on or near the soil surface as the organisms die. Microbes decompose plant and animal residue, as discussed in Chapter 5 and diagrammed in Figure 5–1. During decomposition processes some of the carbon in the organic residue is combined with oxygen in the soil air to form carbon dioxide (CO_2) and vented back to the atmosphere. When the soil is saturated with water and there is no air in the soil pores, the microbes do not function effectively and the organic residues remain only partially decomposed. Under nearly constant saturation, partially decomposed organic matter continues to accumulate on the surface of the soil. During short periods of time when the soil surface is not saturated, the plant remains may partially decompose and lose their shape, but less organic matter is completely decomposed than is added. With time this blanket of organic material thickens and the surface of the soil moves upward.

Soil material containing more than 20% or 35% organic matter (12%–20% organic carbon), depending on clay content, is considered organic soil material. All soils with a layer of organic soil material more than 16 inches (40 cm) thick are classified as Histosols (see the appendix) or simply organic soils, and occupy about 2% of the earth's land surface. Many organic soil materials are nearly 100% organic matter, but some inorganic minerals are present in most organic soil material. Histosols are often called *peat* soils if plant parts can be recognized or *muck* soils if there has been sufficient decomposition that the plant parts are not recognizable. The Anglo-Saxon word *moor,* meaning an extensive area of waste ground overlaid with peat, and usually more or less wet, is often used to identify these areas.

Organic soils develop in areas where the soil is saturated by groundwater close to the surface much of the year. The saturated condition inhibits the amount of oxygen in the soil and retards the microbial decomposition of organic carbon. Cold temperatures also slow microbial decomposition of organic residues, and organic soils tend to be more prevalent but not restricted to colder climates. Saturated conditions typically occur in three topographic locations in the landscape (Figure 6–7). In valley bottoms, usually

FIGURE 6–7 Schematic cross section indicating landscape positions where organic soils often form.

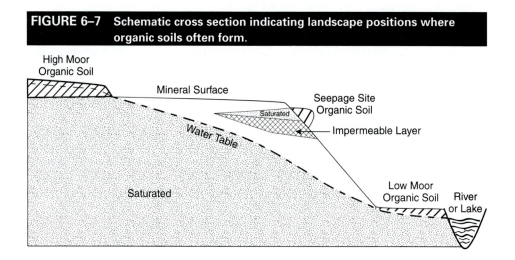

near rivers or lakes, groundwater flows into the valley and saturates the soil much of the year. Organic soils formed in these landscape positions are often called "low moors."

On extremely flat land, with a humid climate and where rainfall exceeds evapo-transpiration much of the year, groundwater saturates the soil much of the year. A common name for organic soils formed in these areas is "high moor." Small areas of organic soils sometimes form on sloping land where groundwater seeps to the surface over impermeable layers.

Low moor organic soils are formed in depressions and saturated by groundwater that has slowly moved through the surrounding soil and geologic mineral material under the soil (Figure 6–8). If the material through which the groundwater has moved is rich in essential elements, the groundwater enriches the low moor site and the organic soils are often very fertile. This is especially true in parts of northern Europe and the United States where glaciers deposited materials from limestone sources during the last ice age.

In high moor landscape positions, the groundwater slowly moves away and slowly removes essential elements. As the organic layer thickens to the point that plant roots do not enter the underlying mineral material, there is no mineral source of nutrients for

FIGURE 6–8   Photo of a low moor organic soil formed at the base of a slope in Wisconsin, USA. Note the presence of water in the trench.

FIGURE 6–9 Photo of an area of organic soil vegetated only with grass on a side slope where water seeps to the surface over an impermeable layer of sediment near Brasilia, Brazil, SA.

the plants growing on these high moor organic soils except in airborne dust. All essential elements have to be obtained from decomposing organic residues. With the slow but nearly continuous export of nutrients in the groundwater flow away from these areas, the organic soils developed become more and more nutrient poor over time. Plants that require only small quantities of those essential elements derived only from inorganic soil minerals begin to dominate the natural ecosystem. Many of these plants are so nutrient poor that grazing animals are not able to find the nutrition they need and populations decrease. Such infertile areas of high moor organic soils are associated with desolate areas in ghost and mystery stories.

Organic soils that form from seepage on side slopes may be either fertile or infertile depending on the mineral composition of the material through which the groundwater is moving. Such areas are usually small but unique (Figure 6–9).

All organic soils are black in color and rich in organic matter, but the nutrient value of the organic material is not seen except with chemical analysis (Box 6-3). Artificial drainage is required to alleviate the saturated conditions and provide aeration for crop production on organic soils. The physically soft and friable consistence of organic soils is prized for vegetable crops, and many areas of organic soils are intensely cultivated. However, when the water table is lowered by artificial drainage, the organic matter slowly oxidizes to carbon dioxide and the surface of the organic soil subsides. Also, when the surface of an organic soil becomes dry during prolonged

BOX 6-3

Beautiful Charlatan

The element carbon is ubiquitous in soils, air, and all forms of life. We recognize that carbon can be as beautiful as a diamond set in a fine piece of jewelry, as dull as the graphite in our lead pencils, or as obnoxious as coal dust. Organic carbon compounds are equally diverse in the role they play in soils. Universally when present they color the soil black. We have learned from experience that most of the naturally fertile soils like those in the highly productive farmlands of the midwestern United States have a distinctly blacker color than many other soils considered less fertile. We associate black color with fertile topsoil we may buy as potting soil for house plants or spread in our flower beds. In most instances our perception that black-colored soil is beautiful is correct, but in some respects soil organic carbon is a charlatan. It can deceive us in several ways and is especially deceptive when it is the major component, as in *organic soils* or *organic soil material*.

Because organic carbon compounds are created in plants, they almost always contain some amount of all the life-essential elements. However, note from Table 4–3 the high C:N ratios present in some organic residues. When organic soil material with a C:N ratio greater than 32:1 is added to soil,

the availability of nitrogen is decreased. In Table 5–3, note how the content of Ca and K in leaves reflects low contents in the soil. We must conclude that all soil organic carbon is not of equal value with respect to the content of essential elements needed for plant growth.

Organic soil material and especially organic soils often deceive us by their inherent low *bulk density* (see Box 2-3). Some organic soils (Histosols; see the appendix) have bulk densities less than 0.1 Mg m^{-3}. Thus all chemical and physical data reported on a weight basis greatly overrepresents the quantity present per unit volume of soil as explored by plant roots. Almost all organic soils have bulk densities that are substantially less than most mineral soils. The cation exchange capacity (CEC) of organic compounds in soil is strongly related to the pH at which CEC is determined (see Box 4-2). The amount of CEC attributed to organic matter is grossly overestimated by methods that determine CEC at pH 7 or pH 8.2 if the pH of organic soil is less than 7.

When evaluating organic soils and organic soil material, it is wise to adhere to the cliché "Don't judge a book by its cover." Organic soil material looks black and beautiful, but it can be as valuable as a diamond or as troublesome as coal dust.

drought, lightning or human-created fire can ignite organic soils. A single fire may burn several inches or more of the organic soil, and fires may burn slowly for several years (Figure 6–10).

Organic soils are mined in many areas of the world and burned for fuel both in household fireplaces and in large power plants to generate electricity (Figure 5–4). In the geologic history of the earth, areas of organic soils have been buried by mineral depositions and compressed into coal, natural gas, and oil.

FIGURE 6–10 Photo showing the loss of approximately 30 inches (76 cm) of organic soil due to one fire in eastern North Carolina, USA.

"CAT CLAY": ACID SULFATE SOILS

Some of the most singular soils in the world are commonly known as *cat clays,* which form in unique conditions present along portions of ocean and other brackish (salty) water shorelines. The term is an English translation of a Dutch expression and appears to have originated in The Netherlands and Germany where such soils were sites of so-called mysterious bad luck in the form of crop failures. The source of the bad luck has been found to be large amounts of sulfur in the soil. Sulfur is present in salty ocean water and combines with iron present in fresh water flowing from the land to form iron sulfide (FeS). Eventually within water-saturated, oxygen-poor soils, the mineral pyrite (FeS_2) is formed (Fanning and Fanning, 1989). These iron and sulfur compounds are not injurious to plants, but when drainage ditches are dug to remove the saturating water and make cultivation possible, air containing oxygen enters the soil voids, sulfuric acid is formed, and the soil pH values drop to 3.5 or lower. It usually takes a few years of drainage for oxidation of the sulfur to create the extremely acid conditions. Few plants are able to root in such acid soil, the crops fail to grow in fields where previous crops were successful, and the farmer has had bad luck. Not all soils along seacoasts are sulfur affected, and it is mainly clayey-textured and organic soils near where freshwater rivers enter an ocean that tend to be most affected.

Cat clays can often be detected by their rotten egg hydrogen sulfide (H_2S) odor or the visible presence of pyrite. Usually a chemical analysis is required to verify that sufficient quantities of sulfur are present to strongly acidify the soil upon drainage. A very small amount of hydrogen sulfide emits a rotten egg odor even if the quantity of sulfur is so low that oxidation will not cause severe problems. If an acid sulfate soil is kept saturated and not allowed to oxidize and form sulfuric acid, plants can grow well on cat clays. Paddy rice has been successful in many fields when farmers are able to keep their fields flooded. Should termites bring some of the sulfur-bearing soil material to the surface in their mounds where it oxidizes, the acid formed often kills the crop in the area surrounding the termite mound.

Severe acid sulfate problems may be created in construction or mining operations when sulfur-bearing materials dug from saturated layers deep below the earth's surface are dumped on the land surface where they oxidize and form sulfuric acid. The spoils from coal mines are frequent sources of sulfate materials. Once oxidation takes place it is extremely difficult to neutralize the acidity, and each site must be carefully studied to determine the best reclamation methodology.

TOO MUCH WATER: "HYDRIC SOILS"

Organic soils and cat clays result from specific conditions of soil saturation. Many other soils experience saturation during some period of time in most years but never accumulate enough organic matter to become true organic soils. Such soils do contain more organic matter than surrounding soils that are seldom saturated. Most commonly these soils are called "poorly drained" or "somewhat poorly drained," and artificial drainage is required for cultivation and the growing of most food crops. Recently the term *hydric soils* has become part of wetland regulations in the United States and can be associated with some but not all of the soils that occasionally become saturated with water at or very near the soil surface. When a soil is saturated and soil microbes have exhausted the oxygen dissolved in the soil water, there are certain microbes that can use oxygen that is already part of other minerals and ions. One of the first ions that can provide oxygen to these microbes in a water-saturated soil is the nitrate ion (NO_3^-). When microbes utilize the oxygen from the nitrate ion, nitrogen gases such as N_2 are formed and escape to the atmosphere. This process is called *denitrification* (see Chapter 4).

In the absence of adequate oxygen (O_2), some soil microbes can obtain oxygen from iron (Fe_2O_3) and manganese (Mn_3O_4) oxides. This process, called *reduction,* causes normally red-, yellow-, or black-colored iron and manganese oxides to dissolve as colorless ions in the soil water that are free to leach from the soil. The absence of red and yellow colors associated with iron oxide coatings on the silicate minerals in the soil leaves the soil with the gray color of silicate clays and sand unless colored black by organic matter.

Poorly drained or hydric soils are present in parts of the landscape in all areas with humid climates and are sometimes present where high water tables and saturated conditions are present in floodplains in arid and semiarid lands. Like many of the processes in soils, saturated conditions followed by denitrification and reduction of iron and manganese oxides are sporadic conditions caused by weather events. Where these conditions can be expected to occur within the landscape is discussed in Chapter 7.

Poorly drained or hydric soils can be identified by continuously monitoring the depth to saturated conditions, but this is laborious and time consuming because the occurrence of saturated conditions is sporadic and weather related. It is more common to identify poorly drained soils by the presence of gray colors in the subsoil that result from the reduction and loss of iron oxides.

Where it is possible to install artificial drainage systems to control the upper limit of saturation, that is, the water table, most poorly drained soils are highly prized for the production of food crops. Artificial drainage systems can take several forms but almost all require digging ditches in which excess water can escape the area during periods of excess rainfall (Figure 6–11). Closely spaced open ditches necessary to remove saturating groundwater hinder the operation of farm equipment. In many locations only a few main ditches are constructed and porous pipes are buried 3 to 5 feet (90–150 cm) below the surface of the land at intervals between the ditches. The interval between the porous pipes is determined by the lateral permeability or texture of the soil. In rapidly permeable coarse-textured (sandy) soils, the interval between the porous pipes is usually between 100 and 150 feet (30 to 45 meters) and in slowly permeable clayey soils, it is 30 to 40 feet (9–12 meters). The porous pipes are in effect permanent large voids in

FIGURE 6–11 Photo of field ditches and formed ridges (bed planting, tillage) in a field of high moor organic soil in eastern North Carolina, USA. The ditches lower the water table, and corn will be planted on the ridges to further protect their roots from saturation. Note flock of Arctic swans gleaning seed from a previous crop in the field.

the soil, and when the soil is saturated, water from the larger voids in the soil flows into them allowing air to enter into otherwise saturated soil. Historically, the porous pipes used were constructed of baked clay or concrete tubes called tile. The practice of installing tile to drain water from naturally saturated land was called *tiling.* More recently, perforated plastic pipes are used. The tile or plastic pipes open into the surface ditches to discharge the water.

The installation of tile creates a greatly improved media for crop growth in frequently saturated soils. The most readily observed benefit of any drainage is improved physical support that allows the mobility of farm machinery that would otherwise become mired in the mud of saturated soil. By preventing the water table from rising to the surface during extremely rainy periods, the crop roots are protected from saturated conditions that are conducive to root rot pathogens. Many crop plants are killed if their root system remains in saturated soil for more than about three days. Tiling is especially beneficial in temperate latitudes with relatively short warm growing seasons because lower water content near the surface soil allows the soil to warm more quickly in the spring and permits earlier planting of crops. If a prolonged dry period develops during the growing season, the saturated conditions that are present immediately below the depth drained by the tile contains not only the water held as available water by the medium-sized pores in the soil but also the water stored in the larger pores (see Chapter 2). This provides an extra source of water that buffers the crop from severe water shortages that occur during droughts on adjacent well-drained land where the water table is below the depth that can be reached by crop roots. Some form of artificial drainage has been installed on millions of acres of farmland throughout the Midwest of the United States and elsewhere in the world (Pevelis, 1987). Almost invariably these poorly drained hydric soils, with properly installed and maintained artificial drainage systems, have higher crop yields per acre than any other soils in the area.

Poorly drained soils are highly prized for growing rice under flooded or paddy management systems. Rice, unlike most crop plants, is able to transmit oxygen from the aboveground portion of the plant to its roots. In flooded or paddy management, farmers seek to maintain water above the soil surface for much or all of the growing season by constructing low mounds of soil (dikes) around their fields. The dikes are constructed to restrict loss of rainfall via runoff and can be used to retain irrigation water if available during prolonged dry weather. In preparation for planting rice, the fields are often *puddled,* which compresses the wet soil by collapsing the large pores, thus restricting the leaching of water. Paddy rice farmers often march animals over the field to puddle the soil by compressing the soil with their hoofs prior to planting rice (Figure 6–12).

The use of so much of the land with poorly drained soils for crop production has caused environmental concerns in the United States, and regulations have been enacted that limit further artificial drainage of poorly drained (hydric) soils. The process of nitrate reduction (denitrification) that takes place when soil is saturated is seen as reducing the nitrogen content of groundwaters used for drinking water and in the water that enters streams, rivers, and lakes, thereby reducing algae and other undesirable plant growth in those surface waters. The presence of hydric soils is a criterion for the jurisdictional definition of wetlands.

FIGURE 6–12   Photo of water buffalo being driven around in a rice paddy to puddle the soil before the planting of rice in Indonesia.

NEW FROM THE TOP DOWN: VOLCANOES AND FLOODPLAINS

Most soil forms as water infiltrates through the land surface where it mixes with organic acids and slowly decomposes, or "weathers," the primary minerals of the underlying mineral pool to form soil material. Because more water moves through the surface soil than through the subsoil, the most weathered or decomposed mineral material is usually in the topsoil, and less weathered mineral material is deep in the subsoil. There are two rather extensive exceptions to this general pattern of soil formation.

During the course of volcanic eruptions, molten ejecta is thrown into the air where it cools and solidifies into ash and falls to the earth. Large and violent eruptions are brought to our attention via the evening news, but less violent eruptions are more frequent around active volcanic vents (Figure 6–13). Very fine particles of ash are blown in the wind and deposited locally on the soil surface in the area around the volcanic vent. Molten volcanic ejecta cool and solidify very rapidly when thrown into the air during volcanic eruptions. The ash material formed is poorly crystallized and rather rapidly weathered to amorphous (short-range-ordered) clay minerals once deposited on the soil.

Although soils formed from volcanic ash occupy less than 1% of the land surface, their uniqueness has led to their identification as one of 12 major kinds of soils

FIGURE 6–13   Photo of smoke and fine ash exhausting from a volcanic cinder cone in Costa Rica, CA.

(i.e., Andisols; see the appendix). Sporadic deposition of volcanic ash adds essential elements to the mineral pool at the soil surface. The relatively rapid weathering of ash material releases essential elements to the available pool providing fertility for plant growth (Figure 6–14). The dense human populations in places like Java, Japan, Central America, and near the Rift Valley in East Africa attest to the sustained food-producing capability of these soils.

In a similar fashion, and more common throughout the world, is the deposition of soil material in floodplains along rivers. The mud left in the houses, streets, and on the fields after a flood event is topsoil and plant residue eroded from the watershed drained by the flooding river. In the United States there are extensive areas of fertile floodplains along the Mississippi, Missouri, and Platte Rivers, and numerous smaller rivers, within which we can observe the effect of this sporadic but predictable addition of soil nutrients to the soil surface. Ancient civilizations along the Nile River, dense human populations on the deltas of the Mekong and Ganges Rivers in Asia, and the floodplains of many other rivers in the world are associated with fertile soil material deposited on the surface of soils by floodwater.

OVERVIEW OF UNIQUE SOILS

In some respects every soil and every parcel of land is unique. Humans can do little to warm soil too cold for crop growth, but early civilizations learned to provide water for food crop production in arid land and drain water from saturated land. The benefits

FIGURE 6–14   Photo of an inactive volcanic crater surrounded by intensely cultivated fields on the foot slopes in Ecuador, SA.

gained from irrigation have served to concentrate humans throughout human history. Humans are slow to recognize the detrimental affects of soluble salt accumulation that follows inadequate leaching of irrigated soils, and bitter disputes over limited supplies of water are omnipresent in arid lands. In like fashion, the artificial or engineered drainage of poorly drained mineral and organic soils has created some of the most productive cropland but created concern for the loss of wetlands in many parts of the world. Acid soil with aluminum-toxic conditions that often forced abandonment of land are routinely ameliorated by additions of lime in modern agricultural production systems but remain as a major problem in many lesser developed areas of the world. Alkali and cat clay conditions were seldom understood by early civilizations, and where present today they pose management challenges. The volcanic and floodplain areas were unique to the development of early civilizations because they naturally renewed the chemical elements necessary for sustained human food production. The role of some of these unique areas in human civilization is discussed more fully in Chapter 8.

LITERATURE CITED

Buol, S. W., and M. L. Stokes. 1997. Soil Profile Alteration Under Long-Term, High-Input Agriculture. In R. J. Buresh, P. A. Sanchez, and F. Calhoun (Eds.), *Replenishing Soil Fertility in Africa* (pp. 97–109). Soil Science Society of America Special Pub. No. 51. Madison, WI.

Fanning, D. S., and M. C. B. Fanning. 1989. *Soil Morphology, Genesis, and Classification.* New York: John Wiley.

Hillel, D. J. 1991. *Out of the Earth.* New York: The Free Press.

Letey, J. 1994. Is Irrigated Agriculture Sustainable? In *Soil and Water Science: Key to Understanding Our Global Environment* (pp. 23–37). Soil Science Society of America Special Pub. No. 41. Madison, WI.

Pevelis, G. A. 1987. Economic survey of farm drainage. In G. A. Pavis (Ed.), *Farm Drainage in the United States: History, Status and Prospects* (pp. 110–136). U.S. Department of Agriculture-Economic Research Service Miscellaneous Pub. No. 1455. Washington, DC.

CHAPTER REVIEW PROBLEMS

1. What percentage of the earth's land area is cultivated for food production? What percentage is estimated as having potential for cultivated food production? Why does the remainder of the land have little or no potential for cultivated food crops?

2. Why does salt accumulate in soils? How can it be removed?

3. How are acid and aluminum-toxic conditions in a soil controlled?

4. Why does organic matter accumulate to form organic soils?

5. What happens to nitrogen in hydric soils? Why?

6. Why are areas of volcanic activity and areas of frequent flooding often the sites of dense human habitation?

CHAPTER

7

Soil Families on the Land

In no area of the world are soils uniform over all the land. However, within most areas there are similarities among the soils, and as a group they often contrast with soils in more distant lands. Soils are difficult to see. It is common practice to refer to soils by their association with more easily seen features. Geography is easy to identify, and many of the more common soil identifications are made on the basis of geographic location. For example, we often speak of tropical soils to identify soils located geographically within the latitudes of $23°, 27'$ north and south of the equator. Soils in the tropics have only one property in common. They have very little temperature change throughout the year in contrast with soils in temperate latitudes that experience greater seasonal temperature changes (Chapter 3). Except for a lack of seasonal temperature changes, most soils in the tropics have the same chemical and physical properties as some soils in temperate latitudes.

Similarly, some people identify soils by climatic zones, such as arid soils, or by the type of natural vegetation that is present. We often speak of forest soils, grassland or range soils, swampy soils, or desert soils. Landscape features such as mountain soils, floodplain soils, or hilly soils often are frequently associated in attempts to identify soils. Sometimes, a combination of identifiers such as prairie grassland soils, swampy marsh soils, or forested mountain soils attempts to communicate soil properties. When land is used for agriculture or other commercial enterprises, we may refer to the soil by the crop grown, such as corn soils, wheat soils, rice soils, timber soils, pasture soils, or urban soils. Identification of soil by such associated features or land use activities is of value only if all the people communicating are familiar with the area being discussed. Terms that identify only location or land use convey little about soil properties.

Some common expressions such as sandy soils, clayey soils, loamy soils, or muck and/or peat soils do relate to soil properties. Textural references usually relate only to the topsoil, although some persons experienced with digging or cultivating soil refer to clay pan soils or shallow rocky soils to express features found in the subsoil.

It is common to refer to the color of the soil as a means of identification. Soil color, often qualitatively modified as good black soils, poor red soils, or sticky gray or blue soils are but some of the expressions. "Good" or "poor," "fertile" or "infertile" are expressions that most often relate experiences people have had in growing specific plants on a particular soil regardless of the reason for their success or failure. The common identification "topsoil" has little or no meaning except that it is the layer of soil stirred by cultivation, usually black in color and *presumed* to be fertile. Although presumably fertile, dark-colored topsoil may be almost any texture, acid or alkaline in reaction, and have essential elemental contents that differ greatly from area to area.

Within farming communities, indigenous expressions have taken on rather specific meanings that may be locally well known. Most areas have indigenous expressions for sandy, droughty, clayey, and wet soils. Some of the expressions are quite colorful. For example, in some areas in the southeastern United States, the expression "dinner-bell soils" locally identifies soils rich in sticky clay. These soils have a very narrow range of moisture content within which they are easily cultivated. The dinner-bell expression infers that the soil may be too wet to cultivate in the morning and too dry to cultivate in the afternoon; thus it must be cultivated only at dinnertime, that is, at noon because the midday meal is traditionally called dinner in those localities. Although this is a bit of an exaggeration, it expresses a practical problem when farming these soils. In some localities in the upper Midwest of the United States, some soils are referred to as "push dirt," indicating that a plow does not slice through the soil but rather the soil material sticks to the plow and pushes ahead of it. "Gumbo" is also an expression used in some communities for soils that have a strong tendency to stick to tillage tools. Some indigenous terms convey very specific information about soil properties, but that information is limited to indigenous people with local experience and has little widespread applicability.

Soils are often identified by type of geologic material from which they form. This type of identification has somewhat universal merit because it helps identify the mineral components in the mineral pool. *Sandstone* soils are most often of sandy texture and low in content of essential elements. *Shale* soils are usually shallow and have loam or clay texture. *Limestone* soils are rich in calcium and often magnesium and have near neutral pH values but are frequently shallow to hard rock. *Granite* soils usually have a particle-sized mixture of sand and clay with few silt-sized particles. *Glacial till* soils are formed in deposits left behind as glaciers melted and consist of a mixture of particle sizes usually including many stones and rocks. *Glacial outwash* soils have layers of sand, gravel, and finer particles reflecting the speed of the water at the time they were deposited. All so-called glacial soils are not equal in mineral composition. Glacial deposits containing substantial amounts of limestone are relatively more fertile than those containing only material from acid igneous rocks. *Loess* soils formed from windblown dust widely deposited over the land in Europe, Asia, and North America as the most recent continental glaciers retreated approximately 10,000 years ago. Loess soils are usually deep, free of rocks and stones, and have a high content of silt-sized particles. Although there are exceptions, most soils formed from loess have many properties in common and are intensively used for crop production in Eastern Europe, Central Asia, and the Midwest of the United States.

Landform identifiers such as *level, rolling, hilly,* or *mountainous* identify the steepness and shape of the land but little about other soil properties. Other commonly used landform expressions include *coastal plain* soils formed on nearly level topography in deposits left as sea level lowered. Soils formed on coastal plains usually contain a variety of materials eroded from distant hills and deposited in the oceans. The term *piedmont soils* usually identifies soils formed from rock on rolling to hilly landforms but offers little identification of specific soil properties unless the rock type is also identified.

All references to soil by association with more easily identified features of geography, climate, land use, topography, or vegetation have limited credence in identifying specific soil properties. They are better considered land identification terms rather than soil identification terms. However, the shape of the land surface is one of the most significant factors in crop production and is a rather reliable indicator of soil properties within localized areas. The spatial association of different soils as related to the shape of the land can be considered a spatial soil family.[1]

THE HILL FAMILY: CREST, SHOULDER, SIDE SLOPE, TOE SLOPE, AND BOTTOMLAND

Within every area of the world there are spatial sequences of different soils related to landscape features. Soil properties are most often related to landscape shape because of water movement over the land surface or the relationship of the land surface to the depth of the water table below the land surface. Water flows downhill over the surface of the soil and erodes soil particles from the higher elevations and either deposits those soil particles on the more level land below or carries them into rivers, lakes, and streams. Many soil differences are determined by the shape of the land surface and related to the distribution of soil particles transported by water flowing over the soil surface.

All hills have distinct components that can be identified by relative positions in the landscape (Figure 7–1). Crest, shoulder, side slope, toe slope, and bottomland are identifiable parts of all hills. Although the relationship of soil properties to landscape position is rather universal, distinct differences related to climatic conditions, the geologic materials present, and the size and shape of the hills preclude universal interpretations of landscape–soil property associations.

Most everyone would agree that when we see a part of the landscape that rises above the surrounding area, we are looking at a hill. Hills come in all shapes and sizes. How high above the surrounding landscape does an area have to be before we call the area a hill? According to my dictionary, a hill is a natural elevation smaller than a mountain and a mountain has greater height than a hill. There is no specific size or shape to define hills and mountains. In extremely level land, local inhabitants may recognize a hill where the elevation increases by only a few inches above the surrounding land, whereas people accustomed to mountains may consider all of the land level.

In humid climates, a water table, that is, a zone of saturation (Chapter 2), is often very close to the soil surface. Soils that have a thickness of nonsaturated soil above the water table great enough for the rooting of most plants are known as *well-drained* soils.

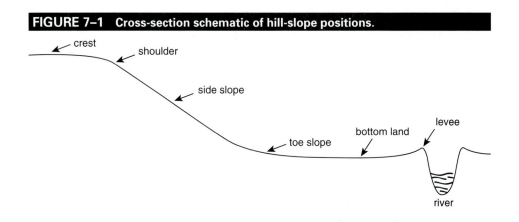

FIGURE 7–1   Cross-section schematic of hill-slope positions.

Soils where the water table is often closer to the surface are known as *poorly drained* soils (Figure 7–2). The terms *well drained* and *poorly drained* are specifically defined in more complete soil classification systems but derive their identity from practical agricultural practices.[2] Poorly drained soils require that drainage systems of ditches or tile lines be installed to remove excess water artificially and effectively lower the water table to facilitate cultivation of food crops. Well-drained soils can be cultivated for food crops without providing drainage systems.

In humid areas where the water table is near the land surface, the size and shape of hills in the landscape is related to the depth to the water table. Where the hills have rounded or convex shape, the thickness of the saturation-free depth of soil is greatest at the highest part or crest of the hill, which usually is present near the midpoint between the rivers and streams that drain surface water from the area and is known as the *interstream divide*. Well-drained soils may also be present under the shoulder and side-slope portions of convex hills. Poorly drained soils are present on the toe slopes and surrounding bottomland (see Figure 7–2a). The flow of groundwater passes through almost all of the poorly drained soils in the landscape positions. As groundwater slowly moves there is denitrification of nitrate and reduction of oxides before the groundwater enters the rivers.

In humid areas where the soils are formed on nearly level landscapes, such as water-deposited coastal plain sediments, the crest of the so-called hill is nearly flat. In these level areas the water table is often near the soil surface near the center of the interstream divide or crest of the hill. Poorly drained soils are present both on the crest of the hill and on the surrounding toe slopes and bottomland. Hydrophytic vegetation grows naturally on most of the poorly drained soils at both locations and a lack of

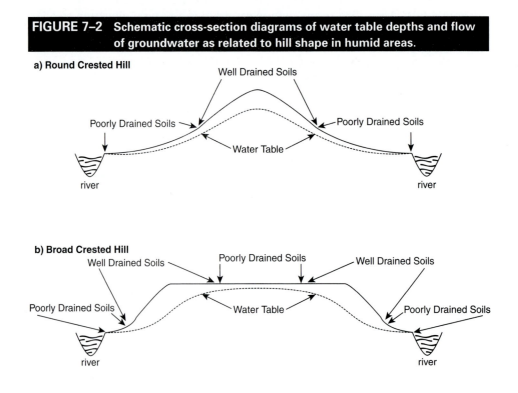

FIGURE 7–2 Schematic cross-section diagrams of water table depths and flow of groundwater as related to hill shape in humid areas.

oxygen engenders reduction of oxides and denitrification. However, the poorly drained soils at the two sites function somewhat differently because of their position in the landscape. The poorly drained soils on the crest positions reduce and denitrify only water infiltrated from precipitation at that site whereas the poorly drained soils near the rivers also reduce and denitrify flowing groundwater that may be enriched with nitrate from the intervening well-drained soils. Well-drained soils are present only on the shoulder and upper side-slope portions of the landscape (see Figure 7–2b).

In arid and most semiarid areas, where the depth to water table often is hundreds of feet below the soil surface, hill slope relationships to soil drainage is seldom applicable, and all of the soils are well drained except in small areas of bottomland near the rivers.

The size and shape of hills has a profound influence on the movement of water over the soil surface during rain events. Runoff water moves soil particles and redistributes them on the surface of the soil. Soil particles are most rapidly removed (eroded) from those parts of the landscape where the volume and velocity of the runoff water is greatest and deposited as the velocity of the runoff water decreases. When rainfall intensity, surface soil texture, and vegetative cover are equal, the amount of runoff is greatest on steeper slopes and less on more level slopes. On the crest of all hills, the slope is less than on the shoulder and side-slope positions; thus erosion rates are less on the crest of a hill than on the shoulder or side slopes below the crest. During some severe and prolonged erosive rainfalls, some runoff water flows into adjacent rivers. However, many and perhaps most rainfall events that create some runoff and erosion on the steepest portions of the landscape are not of sufficient duration or intensity to carry all of the suspended soil material to the rivers. During these events the flow of runoff water is slowed by dense vegetation such as grass, and much of the suspended material is deposited on the soil surface in the less steeply sloping lower portion of the side slope (i.e., the foot slope, toe slope and bottomland positions on the landscape).

In almost all landscapes, the greatest amount of erosion takes place on the shoulder portion of the hill. On the shoulder little sediment derived from the crest of hill is available for deposit because the angle of slope on the crest of the hill above the shoulder is less steep than at the shoulder. The runoff flow of water accelerates at the shoulder and carries any suspended soil material further down slope. The white strip of land above the workers in Figure 7–3 identifies the shoulder of the hill. The white color of the soil surface results because soils on the shoulder are shallow, and erosion is exposing a layer of white-colored, carbonate-rich subsoil.

In landscapes of greater relief, higher hills (mountains) and steeper slopes underlain by hard rock, the shoulder position is often a nearly vertical rock face. In Figure 7–4 the vertical exposures of rock at the shoulder position have resulted as both surface erosion and landslides removed soil from the shoulder and the side slope below exposing hard bedrock at the shoulder position on the slope (Box 7-1). Exposures of vertical rock faces on the shoulder of the hill slope are sometimes referred to as "rim rock."

The bottomland and toe slope positions in the hill family receive soil material eroded from higher elevations on the slope. The foot slope and toe slope are the lower portion of the side slope where the angle of the slope decreases, velocity of runoff waters slows, and some material eroded from the adjacent slopes is deposited. Bottomland is the almost level land below the toe slope. Portions of most bottomland adjacent to rivers are subject to occasional flooding and commonly designated as floodplains. Studies have shown that as much as 90% of the soil material eroded from

FIGURE 7–3 Photo of farmers planting a field in Ecuador, SA. Note the area of white lime accumulations being exposed by erosion on the shoulder of the hill above the workers.

FIGURE 7–4 Photo of a gully formed in a landslide area in Madagascar. Note the vertical rock face exposed on the shoulder of hill in the background as landslides have removed material from side slopes on this tectonically uplifting island.

BOX 7-1

Erosion and Landslides

In a broad sense, *landslides* can be considered as a form of *erosion* and in common usage the two terms are often equated. However, there is a distinct difference. Landslides result from instability below the land surface, whereas erosion is the removal of material from the land surface. In nature the two processes often overlap. Instability, often in the geologic material below the soil on hill slopes, occurs naturally as stream and river channels deepen, causing the hill slopes to become steeper as the streams remove material from the *foot-slope* and *toe-slope* portions of a hill. At some point in time as this occurs, usually during a prolonged period of rainfall, a thick layer of material on the hill slope rapidly slides down the slope and in so doing removes both soil and vegetation. This event exposes earthen material to the erosive action of wind and water, most often as highly visible *gully erosion*. The erosion features observed after this sequence of events are often thought to be the fault of vegetative disruption by human activity, whereas the triggering cause was instability caused by the natural down cutting of rivers and streams and instability of a saturated layer of material below the soil surface.

Landslides are most frequent in steep mountainous terrain where hill slopes are quite steep and fast-flowing rivers and streams rapidly deepen their channels. Geologically rapid uplift of *tectonic* plates is often a factor responsible for rapid stream channel down cutting and landslides on the surrounding slopes in many areas of the world.

the more steeply sloping soil surfaces in hilly land is deposited on the surface of soils in surrounding toe slope and bottomland positions and does not enter the adjacent rivers and streams. In Figure 7–5 we see a black layer in the subsoil of a soil in a toe slope position. The black layer is a former soil surface that has been buried by the deposition of soil material eroded from the sloping land in the background.

The floodplain portion of bottomland is where the land is covered by water during sporadic flood events. During a flood event the waters in the river rise and spread over the surrounding land. As floodwaters spread over the bottomland, velocity decreases as it encounters vegetation and suspended soil material is deposited. The sediments deposited on the floodplain most often are derived from erosion on hills upstream from the site and contain organic residues that chemically enrich the floodplain.

As the level of water in the river subsides, the floodwater on the floodplain finds that its return to the river is impeded by a levee of higher land adjacent to the river. A natural levee forms adjacent to the river as a result of the decreased velocity of water as it overflows the river or stream channel and spreads over the land. Sand-sized particles are the first to be deposited in the levee adjacent to the river. Silt- and clay-sized particles are carried further from the river and eventually deposited as the velocity of the floodwater further decreases. As this scenario is repeated, a pattern of sandy soils on ridges adjacent to the river or stream and finer textured soils on backswamp flats further from the river is developed in floodplains.

Over time a river channel may shift course as it entrenches into the land and leave behind a *meander belt* sequence of sandy levees and finer textured backswamps

FIGURE 7–5 Photo of soil profiles exposed in a trench in Wisconsin, USA. Note the depth to which the original topsoil (black layer) in the *foot-slope* and *toe-slope* positions in the foreground has been buried by lighter colored material eroded from the slope in the background.

(Figure 7–6). Such a sequence of textures creates a pattern of sand and finer textured soils that have a profound influence on crop growth (Figure 7–7). Meandering river channels also deposit sediments of different texture as the speed of the water changes while rounding bends in the river channel. As river water flows around a sharp curve in the channel the flow velocity slows and sand-sized particles are deposited in what are known as point *bars* (Figure 7–8). In future time as the river seeks another course the point bars will be sites of sandy textured soil surrounded by finer textured soils. Such processes are the source of rather contrasting types of soil ofen in both geologically old and recent bottomland.

Rainfall and runoff, with the associated erosion, are and have always been active on all land and have created a predicable spatial pattern of soil properties associated with all hills. Soils on the shoulder positions tend to be less deep and often have more clayey topsoil than soils on any other part of the hill. In mountainous terrain the shoulder position may be a steep cliff of exposed rock. Soils on the side slope tend to be intermediate in depth because they may erode during intense rainfall events but also receive depositions of soil material eroded from the shoulder position in less intense rainfall events. Soils on

FIGURE 7–6   Aerial photo of *meander belt* of sandy *levees* (with tree growth) and *back swamps,* some filled with water, formed as a large river in the Amazon basin in Peru, SA successively changed course.

FIGURE 7–7   Photo of an irrigated cotton field in southern Arizona, USA. The field is on an old floodplain. The sandy textured soils on the former levee (center of photo), now leveled to facilitate *flood irrigation* have considerably poorer cotton growth than finer textured soils on each side.

Photo courtesy of Dr. Jack Stroehlein.

FIGURE 7–8 Aerial photo of sandy point bars (white areas) in a meandering river in the Amazon basin in Peru, SA. Also note the pattern of active and *fallow* slash-and-burn fields adjacent to the river.

the crest of hills are also intermediate in depth and surface properties because they have more level surfaces and experience less erosion than shoulder and side-slope positions and receive no deposits from run-on water. Soils on the toe slope and bottomland tend to be deep, with thick surface topsoil layers composed of surface soil material eroded from the adjacent shoulder and side-slope positions.

Most soils in a hill-slope family have many similar mineralogical properties because they often formed in similar geologic materials. This is not universally correct because in some landscapes the crest of the hills may be of a different and more resistant sedimentary rock overlying a less resistant rock exposed on the side slopes, such as limestone over sandstone. However, soils in the floodplain portion of the bottomland position often are formed in geologic material eroded from hills far upriver and often have mineralogical properties quite dissimilar from other soils in the hill-slope family. This is most often the case along larger rivers where the soils in the floodplain are mainly composed of soil materials eroded from hills in distant headwaters of the watershed that may be quite unlike the material in adjacent hills. Much of the soil material deposited in most floodplains was eroded from surface layers of soil, and it is almost always enriched with organic residues.

MANAGE FIELDS OR SOILS?

In practice, most land areas have been legally divided into units of property by geographic survey techniques unrelated to the natural landscape positions. Property is usually delineated into units of land with rather rectangular dimensions. In much of the United States, property lines are plotted and recorded by land survey techniques that include square sections, one mile (1.6 km) to a side and further aggregated into 36 square mile townships, insofar as a spherical world allows for square units of land. Rather precise rules have been established to accommodate the necessary offsets so flat pieces of paper can portray maps of a round world. The legal division of land into small house lots seldom conforms to natural landscape features, except in some cases along rivers and legal parcels of land usually have straight edges that cross natural landscape features. Almost every unit of property therefore includes more than one kind of landscape and soil. As people attempt to impose a specific use on any unit of property, they encounter more than one kind of soil.

Much like a host inviting a human family for a visit, the owner or manager of a given property must consider how best to entertain each member of the "spatial landscape family" of soils to assure that all have a "good time." The larger the invited family (property), the more diverse individual members (soils and landscapes) are likely to be. The owner or manager of land may adopt different management options for each type of soil or treat all the land the same. If uniform management is employed, some of the soils will most likely be "unhappy" and not perform well. Different management may be attempted for each kind of soil within the property, but this may prove to be incompatible with the equipment available or simply too expensive. A compromise between these two extreme philosophies is most common.

Within farm-sized properties, different management strategies are usually used in an attempt to accommodate the most extreme soil differences. Where slopes are too steep to accommodate machinery, the land is often used to pasture animals or devoted to woodland and wildlife. Even within individual fields, fertilizer and pesticide application rates may be adjusted according to differing soil properties within the field to assure maximum effectiveness of crop growth and minimize risk of environmental pollution. The contemporary expression of this concept is *precision farming*.

Wise real estate developers seek to determine the natural characteristics of the soils on a given property before they subdivide it into individual house lots. Each lot must contain enough area of soil suitable for the essential components of a home site. These may include acceptable areas for a well and septic system should the area not be on a central water and sewage system, an easily accessible driveway, and reasonable areas for lawns, gardens, and landscaping features desired by a potential home owner. Although all soils can be modified to some extent without making them unhappy, as it were, the most successful urban developments place essential components of a property on soil best suited to a specific function.

WATER MANAGEMENT FOR THE HILL-SLOPE FAMILY

All soils within a hill-slope family are affected by what happens to the other soils in the family. The most significant relationship to remember in this regard is that "water flows downhill." The movement of water is the most common and frequent interaction between members of a hill-slope soil family. As discussed in Chapter 2, the surface of the soil is a major crossroad in the hydrologic cycle. Most of the water from rain or melting snow either infiltrates the soil or runs off over the soil surface. The proportion of water that takes one or the other path is not a fixed quantity for any soil. Slope is the most influential feature controlling the *relative* amounts of runoff and infiltration. The physical condition and vegetative cover greatly influences the proportion of infiltration or runoff that will occur on sloping land. In suburban subdivisions without central storm water conduits, home owners may suddenly find excess water in their lawn when a new house is constructed on an adjacent lot further up the side slope. During construction it is likely that some vegetation was removed and the soil surface compacted by traffic. Also a portion of the new lot is devoted to a driveway, and the house is certain to have a roof. Both allow no infiltration and contribute to increased amounts of runoff.

The processes responsible for creating differences among the soils in hill-slope families are most active during sporadic weather events. Many years may pass between extreme weather events that produce excess runoff water. Although management of land requires application of technology that best utilizes average conditions, land management must include practices that avoid both economic loss and environmental damage that result from extreme weather events. Various techniques have been developed and utilized by humans to affect run-off and infiltration on sloping land. These techniques can be grouped into two broad categories. One category includes techniques that change the shape of the land. The other category includes techniques that control vegetative cover on the land. A combination of the two techniques are often used.

Terraces

Throughout the world people long ago learned how to reduce surface runoff by changing the slope of the land. This is most striking in much of the steep land in Asia where people have constructed level bench terraces and grown rice as a primary food crop (Figure 7–9). By constructing canals to convey water to and from each of the individual terraces, they are able to provide and retain water on each of the terraces. In some areas, dams are built to create lakes in valleys above the terraced land, and water is then available to irrigate crops during rainless periods. Construction of such bench terraces requires thick soils of fine enough texture to support the vertical steps when wet (Figure 7–10). Sandy soils or shallow soils over hard rock cannot be managed in such a way. Clearly, the small size of most bench terraces is only compatible with intense hand labor and not suitable for modern mechanization. Where slopes are less steep than those pictured in Figures 7–9 and 7–10, each terrace step can be wider and perhaps accommodate small mechanized tillage equipment.

The principle of constructing terraces is to modify the natural slope of the land into a series of more level sites and thereby retain water on the soil surface for a longer period of time. This increases the amount of infiltration and decreases runoff.

Strip Cropping

Strip cropping is a commonly used cropland management system for mechanized agriculture on side slopes. Strip cropping provides for the operation of soil tillage equipment and the planting of different crops in parallel bands perpendicular to the slope of the land (Figure 7–11). Most tillage equipment, such as plows and disks, leave small ridges or valleys on the soil surface parallel to the direction of travel. Having these small ridges and valleys perpendicular to the natural downhill flow of water slows runoff and reduces erosion. A primary function of strip cropping is the planting of different crops on alternating strips on the hillside. Each crop is planted and harvested at a somewhat different time each year. By planting different crops in adjacent parallel strips perpendicular to the slope of the hill, the entire slope is not tilled and thus without vegetation

FIGURE 7–9 Photo of narrow bench terraces constructed to control water for growing paddy rice on steep slopes in Sri Lanka.

at the same time. If an intense rainfall event occurs during the time the soil in some of the strips is devoid of vegetation, the velocity of the runoff water from that strip is slowed in the vegetated strip below and total erosion from the side slope is decreased.

Grass Waterways

The best erosion protection of any soil is a permanent vegetative cover of grass with a dense mat of fibrous roots. A dense cover of grass protects the soil from both the impact of raindrops and the eroding energy of flowing water. Establishing a permanent cover of grass in the natural down-slope waterways on a side slope and never cultivating that area can channel runoff water down the slope with minimum erosion. Grass waterways in conjunction with strip cropping are widely used on modern mechanized farms to reduce erosion on hill slopes (Figure 7–12).

Permanent grass-covered waterways are troublesome for farmers because they must interrupt mechanized tillage operations and pass their cultivation implements

FIGURE 7–10 Photo of a tillage operation at the base of newly constructed bench terraces in Indonesia.

FIGURE 7–11 Photo of strip crops for mechanized cultivation in southern Wisconsin, USA.

FIGURE 7–12   Photo of a permanently grassed waterway maintained to allow runoff on a hill slope with strip crops of corn and oats (harvested) in southern Wisconsin, USA.

over a grass waterway. The alternative in many cases is the formation of a gully or ditch that may become impassable, and most farmers quickly realize that the inconvenience of a grass waterway is small compared to the inconvenience of a gully.

Riparian Buffers

Areas adjacent to rivers are known as *riparian zones*. These bottomland or toe-slope positions in the landscape afford a final defense where soil particles eroded from the uplands can be arrested before they enter the rivers and streams and form sediment in lakes and ponds. Maintaining a narrow band of permanent vegetative cover, a few tens of feet wide, adjacent to a river slows the velocity of the runoff water as it passes through the vegetation and suspended particles settle on the surface of the soil rather than being carried into the river. The narrow bands of permanent vegetation in the riparian zone are known as a *riparian buffer* because they buffer the river from sediment (Figure 7–13).

Riparian buffers also can reduce nitrate content of groundwater by the process of denitrification *if* the water table under the buffer is very near the soil surface, a condition often present near a river or stream (Figure 7–2). For high crop yields, it is necessary to have a high concentration of nitrate in the soil within the rooting volume of crop plants. To provide this high concentration, farmers add manure or fertilizer. However, clean water for drinking should contain no more than 10 parts per million (ppm) of nitrate according to present regulations, which is only about 30% of the concentration needed in the water around the roots of crop plants to ensure efficient growth. It has been determined that only a narrow area of saturated soil near the river or stream effectively

FIGURE 7–13 Aerial photo of pastured fields in the Amazon basin of Brazil, SA. Note riparian buffer strips of uncut trees along natural streams.

reduces the nitrate content of groundwater that flows through the soil under adjacent fields before it enters the river or stream (Osmond, Gilliam, and Evans, 2002). In areas where the water table is many feet below the soil surface, the denitrification function of the riparian buffer is not operational.

Till-No Till

Tillage is the mechanical disturbance of the soil surface. Methods of tilling soil range from large plows to hand hoes. Tillage of the soil before a food crop is planted is conducted for a number of reasons. Tillage is desirable to incorporate manure, fertilizer, and lime into the soil physically so nutrients are available to crop roots below the soil surface and less susceptible to being eroded. Tillage retards weed growth, breaks up or buries undesirable plant residue that interferes with planting equipment, and temporally loosens the surface of the soil to increase infiltration of water. Most tillage practices conducted prior to planting leave the soil surface bare of vegetation and thus exposed to the impact of raindrops and erosion. When intense rain falls on bare soil, the surface of the soil is compacted, and on sloping land runoff increases and maximum erosion occurs. A day or two of drying after an intense rain event, the compacted soil surface may form a crust so strong that young seedlings are not able to emerge and additional tillage may be necessary to break the crust. Intense rainstorms cannot be anticipated, and a single such storm that occurs when the soil surface is bare often creates much more erosion on sloping land than the site would otherwise experience in several years.

The development of chemicals that can kill undesirable weeds and genetically altered seed that allows more extensive use of such chemicals have provided alternatives to tillage for weed control. The development of large planting equipment that can uniformly plant seeds through a cover of plant residue where smaller planting equipment could not function makes it possible to plant without tilling the soil. A combination of chemical weed control and modern planting equipment has made it possible to avoid exposing bare soil by planting directly into the residue left by the previous crop, called *no-till farming.*

A major advantage of no-till farming is the use of less energy in tillage operations and thus a lower cost of planting. By leaving plant residue on the soil surface, the danger of crusting is reduced and infiltration rate is increased. On sloping land, runoff and erosion are reduced. Disadvantages of no-till farming are increased use of chemicals to kill the weeds and delayed warming of the soil in the springtime. No-till operations also must carefully avoid surface compaction by not operating equipment when the soil is too wet. It also is difficult to mix fertilizer and lime into the soil without exposing bare soil. Fertilizer placed on the surface is more susceptible to erosion. Also, nutrient elements, such as phosphorus, that do not easily move downward in the soil, are concentrated very near the surface where they may become unavailable to plants during dry weather.

No-till planting became a viable management option on cropland in the United States during the 1990s and by 2002, 36.6% of the cropland was planted with some form of no-till, popularly referred to as *conservation tillage.* In the same year, 40.6% of the cropland was intensively tilled. The remaining 22.8% of the cropland was planted with some form of *reduced tillage,* a compromise between no-till and intense till and usually involves some tillage operations that the farmer deems necessary to break up compacted soil and/or to incorporate fertilizer and lime into the soil (Conservation Technology Information Center, 2002).

Each of the practices just described has numerous variations in design to best accommodate specific landscapes, kind of soil, and management objectives of the farmer. Like all soil management practices, there is no one best practice for all soils. Protecting the bare soil surface from the direct impact of rain is germane to all successful erosion control practices. Practices that increase infiltration and decrease runoff are usually desirable on side slopes but may be of little or no value on nearly level toe slopes, bottomland, and crest positions.

CONTROLLING EROSION?

The exact proportion of the eroded material that is accumulated on the toe slope and bottomland and not transported to the rivers and streams varies greatly with such factors as density of vegetative cover, degree of slope, distances involved, and intensity of rain events. Table 7–1 summarizes the long-term data collected in one intensely studied watershed to illustrate the distribution of eroded soil material within that watershed (Trimble, 1999).

The Coon Creek watershed is 88,920 acres (360 km^2) and located in west central Wisconsin, USA. The watershed empties into the Mississippi River south of La Crosse, Wisconsin. With the national concern for erosion control in the 1930s, Coon Creek watershed was selected as a representative study site to measure the impact of various erosion control practices on the rather hilly farmland in the upper Mississippi River basin.

Beginning in the 1930s, soil conservation practices such as strip crops and terraces were established on the farms within the watershed, and during successive years many

TABLE 7–1 Average Annual Movement of Soil Material in Coon Creek Watershed in Three Eras*

	Erosion Sites			Deposition Sites		
Time Period	Upland Sheet Erosion	Upland Gullies	Tributaries	Tributary Floodplains	Main Stream Floodplain	Mississippi River
1853–1938	359†	80	46	138	309	42
1938–1975	126	71	68	42	183	40
1975–1993	84	21	23	28	61	41

*Upland sheet erosion was calculated from measurements of sediment accumulation at deposition sites and sediment discharge in stream flow.

†All units are 10,000 tons per year.

Source: Trimble, 1999.

farming practices in the area changed. By the period 1975 to 1993, these changes resulted in nearly a 75% decrease in the amount of soil material eroded by sheet erosion (surface flow) from cultivated fields, gullies, and the stream banks of small tributaries in the watershed. The greatest reduction in erosion was in the amount of soil material removed as sheet erosion from farmers' fields. During the same period of time, the annual amount of soil material deposited on the floodplains also greatly decreased. However, the average annual discharge of sediment into the Mississippi River did not significantly change.

This apparent assault on common sense can be understood if we examine the dynamics of erosive events in Coon Creek and similar watersheds. Erosive events are usually one- or two-day weather events of excessive rainfall usually during a period of time when cultivated fields are not vegetated, that is, during a period of a few weeks after cultivation and planting in the spring or in the fall of the year if the land has been plowed after the crop is harvested. Erosive events invariably result in extensive flooding on the floodplain portion of the landscape, and the majority of the suspended soil material eroded from the hill sides is deposited as sediment as the floodwaters spread over the nearly level bottomland adjacent to the rivers. Certainly during the event there is an increase in the amount of material carried to the main river channel, but many years may pass without a severe erosive event. During those intervening years, the main source of sediment carried to the Mississippi River from the Coon Creek watershed is material eroded from stream banks during periods of rapid stream flow caused by moderate rainfall events.

Within watersheds with broad floodplains as components of the landscape, the amount of material eroded from farmers' fields during sporadic erosive events appears to have little effect on the average annual discharge of sediment to the major rivers (Trimble, 1999). This is not the case in steeply sloping watersheds where toe slopes and floodplains are minimal in size or absent. In such landscapes the runoff waters during erosive events are confined to narrow channels, and suspended soil particles from the hillsides are carried directly into the larger rivers, lakes, and ponds.

It is never possible to stop soil erosion completely on sloping land. As long as water flows downhill, the shoulder and side-slope soils will lose material from their surfaces and the soils on the toe slope and bottomland will accumulate soil material on their surfaces. In watersheds where the streams have steep gradients and very narrow floodplains, it is almost certain that lesser amounts of material eroded from the soil surfaces on the hillside will be retained on floodplains within that watershed. Each watershed area has unique and individual landscape features that render the extrapolation of quantitative relationships between erosion and sediment export via the river systems problematic.

DOES SOIL BENEFIT OR LOSE FROM EROSION?

When the question that heads this section is asked in the singular context of all soil, the answer clearly is "both." The natural chemical fertility humans find in the soils formed in deltas and floodplains results from the deposition of fertile topsoil eroded from the soils on shoulder and side-slope positions on adjacent or distant uplands. Excessive erosion from soils on the shoulder and side slopes is clearly a loss and of concern for the future potential of those soils to serve human food production. The potential harm that can be expected from erosion cannot be answered unless the properties of individual soils are examined. Some soils are able to withstand severe and prolonged erosion rates and lose little of their ability to grow plants, whereas the production potential of other kinds of soil is severely reduced by erosion.

Pierce et al. (1983) addressed the question of potential productivity loss due to erosion on specific kinds of soils and their location in the landscape. They developed a computer model that simulated changes in five soil properties: plant-available water, bulk density, aeration, pH, and electrical conductivity (salt content) known to affect root growth and subject to change as a soil erodes. The model assumed that present farming practices of fertilization would continue and therefore essential nutrient supply would not limit plant growth. The model calculated a "productivity index" that reflected the sufficiency of the soil to provide a favorable rooting volume for crop plants as various amounts of soil was lost from the surface via erosion. They selected soils in Major Land Resource Area (MLRA) 105 in southeastern Minnesota, USA, and simulated productivity index values for 25, 50, and 100 years into the future if the soil surface eroded at rates known to occur in that area (Table 7–2).

Their simulations indicated that slope of the land was most the influential factor in evaluating potential loss of productivity index due to erosion. The majority of the land, about 67% of the cropped acreage in that area on slopes less than 6%, would only lose 2% or less of its productivity index in 100 years if previously recorded erosion rates continued. Approximately 25% of the cropland was on 6% to 12% slopes and

TABLE 7–2   Estimated Change in Production Index of Crop Land in Southeastern Minnesota after 25, 50, and 100 Years of Erosion

Slope%	Acreage (1,000s) Acres	Average Loss in Production Index %		
		25 Years	50 Years	100 Years
0–2	140	0	0	1
2–6	471	1	1	2
6–12	225	2	3	5
12–20	56	10	14	15
20–45	21	20	35	56

Source: Pierce et al., 1983.

could be expected to lose 5% of its production index in 100 years if then present erosion rates continued. The majority of the potential productivity index loss would be on the remaining 8% of the cropland with slopes greater than 12%.

Although slope of the land is a major factor in erosion, the specific properties of individual soils must be considered to determine loss of production potential due to erosion. The calculations of Pierce et al. (1983) indicated that a deep soil of uniform silt loam texture on steep slopes with an erosion rate of 33.9 tons $Ac^{-1} Yr^{-1}$ (76 metric tons $ha^{-1} Yr^{-1}$) would lose only 3% of its production index over 100 years. At that rate of erosion, about 40 inches (1 meter) of soil would have been lost, but the composition and properties of the resulting soil would be favorable to continued crop production. Similar soils formed in thick loess deposits have supported civilization for several centuries in the heavily eroded watershed of the Yellow River in the People's Republic of China. In Figure 7–14, a field has been recently formed by leveling land exposed by severe erosion of soil in thick loess deposits. The evidence of past erosion is seen by the presence of a gully wall in the background.

In contrast, shallow soils over rock on lesser slopes with erosion rates of only 11.1 tons $Ac^{-1} Yr^{-1}$ (25 metric tons $ha^{-1} Yr^{-1}$) were predicted to lose 20% of their production potential in 100 years as about 10 inches (25 cm) of soil was eroded (Pierce et al., 1983). Clearly, both slope of the land and the properties of individual kinds of soil must be considered when attempting to evaluate the potential effect of erosion on future productivity.

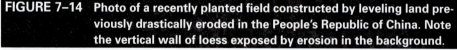

FIGURE 7–14 Photo of a recently planted field constructed by leveling land previously drastically eroded in the People's Republic of China. Note the vertical wall of loess exposed by erosion in the background.

As a point of reference, the erosion rates cited by Pierce et al. (1983) are much greater than average erosion rates in the same area reported to have been 6.3 tons $Ac^{-1} Yr^{-1}$ (14.2 metric tons $ha^{-1} Yr^{-1}$) in 1992 (Argabright et al., 1996). As erosion control practices have been applied, a continuing decline in erosion from cropland is evident. In Iowa, USA, the estimated average annual water erosion from all cropland in 1997 was 4.7 tons $Ac^{-1} Yr^{-1}$ (10.6 metric tons $ha^{-1} Yr^{-1}$), down from 9.9 tons $Ac^{-1} Yr^{-1}$ (22.3 metric tons $ha^{-1} Yr^{-1}$) in 1977 (Iowa Estimated Average Annual Sheet & Rill [Water] Erosion, 2004).

A high proportion of the highly productive food producing land is nearly level or on slopes where erosion hazard is minimal both in severity and frequency. Where small areas of steep slopes are intimately mixed with more level land and therefore included in mechanized cultivation of fields, the steeper slopes frequently have inferior crop yields. Studies in such fields in a piedmont area of the southeastern United States revealed that yield losses on the steeper slopes were largely due to a lack of water because of rapid runoff during summer rains. Because the areas with steeper slopes were small in proportion to the total area of the field, they had little influence on total yield (Daniels et al., 1989).

Numerous studies have been conducted to measure the amount of soil material lost from specific kinds of soils, various conditions of vegetative cover, and landscape positions subjected to measured rainfall events. There can be no doubt that both increased runoff and erosion are directly related to increased slope of the land and a lack of dense vegetative cover. These studies have been instrumental in designing land use practices that reduce the risk of erosion. However, erosion is a natural process that takes place both under natural conditions and under the influence of human cultivation. Natural erosion has resulted in the formation of soils that are shallower on sloping land than on adjacent more level land. Soils formed on toe-slope and bottomland positions that have received sediment eroded from the adjacent slopes are naturally thicker than soils on adjacent steeper slopes.

The question "How much erosion has been caused by humans?" has plagued scientists for generations. How do you measure soil that is not there? Obviously, that cannot be done. Human definitions of eroded soil are therefore written in terms of the properties of the soil that remains. If the top layers of the soil are thin such that cultivation practices expose subsoil material and suitability for use and management is judged to be adversely affected, a soil is identified as "eroded" in the United States (Soil Survey Division Staff, 1993). This designation of eroded may reflect the natural thickness of soil and not correctly identify erosion caused by human activity. One study in the piedmont area of the southeastern part of the United States located a virgin tract of land that was well documented as never cultivated or disturbed by timber harvested. Soils on slopes greater than 5% occupied 60% of the area and were thin enough that subsoil material would be incorporated the first time it was plowed to a depth of 12 inches (25 cm), and the soil would be identified as slightly or moderately eroded according to standard definitions (McCracken, Daniels, and Fulcher, 1989).

Although erosion is a natural process on all sloping land, the rate of erosion can be greatly accelerated when humans remove the protective cover afforded the soil by vegetation. Many cultivation practices are available to increase water infiltration on steep slopes and in so doing reduce erosion. All erosion entails a loss of essential elements as surface soil erodes, but these elements are a potential benefit to the floodplain where the sediment is deposited. Nutrient-rich sediments also nourish aquatic vegetation and life-forms that rely on that vegetation but pose potential detriment for much aquatic life if an abundance of essential nutrients are introduced. Increased discharge of nitrogen and phosphorus into

FIGURE 7–15   **Photo of dead fish on the Neuse River shore in eastern North Carolina, USA in late summer 1968.**

rivers, lake, and oceans increases the growth rate of phytoplankton and other aquatic plants. When this occurs they consume increased quantities of dissolved oxygen from the water creating problems for fish and other aquatic animals that depend on water well supplied with dissolved oxygen. Increased quantities of nitrogen and phosphorus in surface waters can be related to several sources such as urban sewers and industrial waste disposal but certainly erosion from nutrient rich soil contributes to the problem. Often oxygen depletion is seasonal as in the dead zone occurring in the Gulf of Mexico at the mouth of the Mississippi river where each spring oxygen depletion occurs following peak flow related to snow melt and spring rains in the watershed (DeVore, 2001). In other areas lack of oxygen in rivers late in the summer when the flow of river water is slow contributes to the death of fish (Figure 7–15). There are many faces to the process of erosion.

ECCENTRIC SIBLINGS

Thus far the more common hill-slope families of soils have been discussed. There are some members of the total soil family tree that may be referred to as "eccentric." Some soils have rather extreme properties not easily related to the slope of the land that must be considered.

Soils with a high content of montmorillonite clay, classified as Vertisols (see the appendix), expand and become almost impermeable to infiltration when wet. When dry they develop cracks that may be several inches wide and tens of inches deep. When rain falls on dry Vertisols, almost all of the water infiltrates through the large cracks

FIGURE 7–16   Photo of cracks formed to the surface of a Vertisol during the dry season in Zambia.

(Figure 7–16) and moves soil particles from the surface deep into the soil. After being wetted, the clay expands, the cracks close, and the surface of the soil is pushed upward forming a pattern known as *gilgai*. This sequence of events is represented in Box 7-2 After the clay has expanded the soil will infiltrate very little water and most rainwater will run off even on nearly level land or remain on the surface in an interesting pattern of puddles caused by the upheaval of the surface (Figure 7–17).

The extreme conditions of shrinking when dry and swelling when wet present unique challenges for management. The physical pressures created as these soils wet and dry are great enough to move the foundations of buildings. It is nearly impossible to assure that the soil around the foundation of a house will always remain dry. Therefore builders have successfully installed subsurface *drip irrigation* systems around house foundations to always keep the soil wet and avoid the pressures that result as from the change in volume as the soil alternately wets and dries. The inability of Vertisols and other soils with a high content of expanding clay to transmit water when wet precludes their use for septic waste disposal. Almost invariably, Vertisols and similar soils with a high content of montmorillonite clay have good natural chemical properties, but their sticky nature when wet bedevils the operation of tillage equipment and thus strict moisture control is critical.

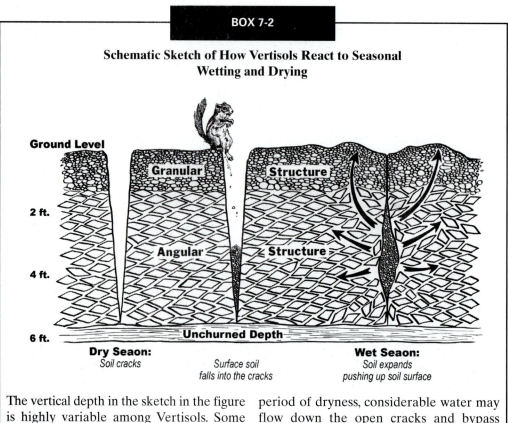

BOX 7-2

Schematic Sketch of How Vertisols React to Seasonal Wetting and Drying

Ground Level

Granular   Structure

2 ft.

Angular   Structure

4 ft.

6 ft.

Unchurned Depth

Dry Seaon:
Soil cracks   *Surface soil*   **Wet Seaon:**
                                 falls into the cracks   *Soil expands*
                                                                  pushing up soil surface

The vertical depth in the sketch in the figure is highly variable among Vertisols. Some Vertisols wet and dry more than once a year, and in arid areas wetting may occur only after several years of dryness. Following a period of dryness, considerable water may flow down the open cracks and bypass (*bypass flow*) much of the soil material. It usually takes several days of wetness before Vertisols fully expand.

Where it is possible to maintain constant wet conditions via irrigation, they can be used for flooded (paddy) rice production, although mechanization is often difficult because of their sticky nature. The unique dynamics of Vertisols and the bedevilment they create to structures and roads needs no exaggeration. However, in Texas, USA there are large areas of such soils and Texas pride causes some Texans brag they have the some Vertisols with cracks that are big enough to swallow cows, jeeps, and cowboys, horse and all.

Extremely sandy soils also present unique features. Unlike Vertisols, sandy soils maintain a high infiltration rate unless they become saturated, and runoff from sloping land is less of a concern for erosion than from loam- or clay-textured soils. However, sandy soils retain only a small amount of available water and require small amounts of natural rainfall or irrigation water at frequent intervals to sustain good vegetative growth. Natural vegetation is often very sparse on very sandy soils even in climatically humid areas. Without a protective cover of vegetation, wind can dislodge sand particles from the easily desiccated soil surface, causing them to bounce along the soil surface, often shearing off the seedlings of crop plants, and accumulate as sand dunes in an area

FIGURE 7–17   Photo of a pasture on Vertisol soils in Texas, USA. Note the pattern of raised areas and water standing in the depressions after a rain when the soil is wet, the clay has expanded, and the cracks are closed.

Photo courtesy of Dr. Larry Wilding.

protected from the direct force of the wind. Small patches of vegetation growing in sandy soils often provide that protected area. Upon encountering the vegetation, the blowing sand particles stop moving and form sand dunes that cover the vegetation (Figure 7–18). Such events may take place more frequently in arid climates, but wind erosion and dune formation is also common on sandy soils in more humid areas (Figures 7–19 and 7–20).

Although sandy soil is the most susceptible to wind erosion, regardless of slope, and clay-textured soils are seldom susceptible because they retain so much water in the very small voids that the clay particles stick to each other, most other soils are vulnerable to wind erosion when dry. In areas frequently subjected to wind erosion, farmers often plant their crops in parallel strips perpendicular to the direction of the winds most likely to cause erosion (Figure 7–21). By having alternate bands of different crops, each planted at different times, strips that are barren are protected from erosive wind by vegetation in adjoining strips. In some places rows of low-growing trees and shrubs have been planted between fields to slow the velocity of the wind near the land surface and reduce wind erosion.

SPORADIC EVENTS ON THE LAND

Erosive rainstorms and floods are sporadic events, but they occur at such frequent intervals of time that almost everyone has some awareness of their impact on the land and human activities. Many natural events that drastically alter the shape of the land occur so infrequently and/or are so localized that they easily escape human recognition. On

FIGURE 7–18 Photo of sand dunes in northern Arizona, USA.

FIGURE 7–19 Photo of sand dunes in Wisconsin, USA.

FIGURE 7–20 Photo of sand dunes along the Outer Banks in North Carolina, USA. Note the effort to control the blowing sand with a slatted fence.

FIGURE 7–21 Aerial photo of crops planted in parallel strips to slow wind erosion in Montana, USA.

FIGURE 7–22 Photo of an area on a mountainside in Bolivia, SA, where a recent landslide has removed all of vegetation and soil, exposing bare rock.

steep slopes, some soils overlay hard rock that do not allow water to move downward. The soil may rapidly infiltrate a limited amount of water, but when the soil over the rock becomes saturated with water during prolonged periods of rain the entire soil slides down the hill (i.e., landslides) (Figure 7–22).

Seemingly solid soils may rest on limestone rock that is slowly dissolving as water flows through fissures within the rock, forming underground caves. As a cave enlarges it may collapse and create a sinkhole in level land. In other areas underground layers of coal may burn causing collapse of the land (Figure 7–23). Tectonic events (earthquakes) drastically alter landscapes, creating unstable hillsides that rapidly erode in some areas of the world.

Floodplains may be present and accumulate sediment for many years. Although they appear permanent to human observation, on occasions of extreme flooding portions of seemingly permanent floodplains may disappear and be replaced by a river channel.

Landslides, earthquakes, sinkhole formation, sand dune formation and the complete removal of land areas in floodplains take place only at sporadic intervals of time. During intervening intervals the land surface may appear stable. Humans are sometimes amazed and consider such events as catastrophic, but they are natural events that shape landscapes and characteristics of some soils in the family of soils on the land.

PERSPECTIVE

With few exceptions, all soils are intimately associated and influenced by their spatial relationship to other soils in the landscape. Movement of groundwater beneath the soil and movement of material to and from land surfaces have shaped

FIGURE 7–23   Photo of a sinkhole being formed in western North Dakota, USA, where an underground layer of coal is burning, causing the land surface to collapse.

landscapes and continue to determine many soil properties on those landscapes. As humans attempt to gain sustenance from the land, their activities affect not only the area they occupy but also adjacent land. Just as a neglected lawn affects the reputation of a neighborhood, unwise land use affects surrounding land and often more distant lands.

No two land areas are exactly alike, but certain simple principles are widely applicable. The first principle is that water flows downhill. Whether that water is flowing over the land surface or as groundwater, anything placed in that water comes to rest someplace in the landscape or adjacent bodies of water. The second principle is that events that shape landscapes are most often spasmodic and often identified as catastrophic. Years may pass without a severe erosive rainstorm that reshapes floodplains or prolonged rain events that trigger landslides that reshape landscapes in mountainous areas. Efforts to portray processes like erosion in human time scales, such as rates per year, fail to capture the dynamics of the process. Perhaps the third principle that applies to all landscapes is that surface water flows faster on steep slopes and ceases to flow where there is no vertical gradient. Humans seek to retain and infiltrate water on sloping land by restricting its path with terraces and strip cropping. Slow-moving water erodes less soil material than fast-moving water, and slowing the speed of runoff water reduces erosion. Erosion of soil material from sloping land is a natural process that can never be completely halted. A permanent cover of vegetation is the best method of reducing the rate of erosion. A permanent vegetative cover is not possible on land where humans plant and harvest their food crops, but several management techniques are available to reduce the exposure of the soil surface in cropland.

Every area of land has some differences related to the slope of the land. Landscape shape is the probably the most visible component of soil and land relationships to life. Although difficult for humans to conceive during their relatively short lifetimes, the shape of the land is not permanent. Although humans can alter some landscape features to reduce undesirable effects of water and wind on the land surface, extreme sporadic and catastrophic events beyond the control of humans naturally occur and drastically alter the shape of the land and the properties of the soil.

LITERATURE CITED

Argabright, M. C., R. G. Cronshey, J. D. Helms, G. A. Pavelis, and H. R. Sinclair, Jr. 1996. *Historical Changes in Soil Erosion, 1930–1992: The Northern Mississippi Valley Loess Hills.* Historical Notes No. 5. Natural Resources Conservation Service and Economic Research Service. Washington, DC: U.S. Department of Agriculture (p. 92).

Conservation Technology Information Center. 2002. National Crop Residue Management Survey 2002. Accessed at http://www.ctic.purdue. edu/Core4/CT/CTSurvey/NationalData.html

Daniels, R. B., J. W. Gilliam, D. K. Cassel, and L. A. Nelson. 1989. Soil Erosion Has Limited Effect on Field Scale Crop Productivity in the Southern Piedmont. *Soil Science Society of America Journal* 53:917–920.

DeVore, B. 2001. Dead Zone Puzzle. http://www. dnr.statemn.us/volunteer/julaug01/hypoxia.html

Iowa Estimated Average Annual Sheet & Rill (Water) Erosion. Accessed at http://extension. agron.iastat.edu/soils/pdfs/WaterErosionChart. pdf 2004. U.S. Department of Agriculture, National Resources Inventory, Agronomy Department, Iowa State University.

McCracken, R. J., R. B. Daniels, and W. E. Fulcher. 1989. Undisturbed Soils, Landscapes, and Vegetation in a North Carolina Piedmont Virgin Forest. *Soil Science Society of America Journal* 53:1146–1152.

Osmond, D. L., J. W. Gilliam, and R. O. Evans. 2002. *Riparian Buffers and Controlled Drainage to Reduce Agricultural Nonpoint Source Pollution.* North Carolina Agricultural Research Service Technical Bulletin No. 318, North Carolina State University, Raleigh. (Also available at http://www.soil.ncsu.edu/lockers/ Osmond_D/web/RiparianBuffers.pdf.)

Pierce, F. J., W. E. Larson, R. H. Dowdy, and W.-A. P. Graham. 1983. Productivity of Soils: Assessing Long-Term Changes Due to Erosion. *Journal of Soil and Water Conservation* 38:39–44.

Soil Survey Division Staff. 1993. *Soil Survey Manual.* Handbook No. 18. U.S. Department of Agriculture (p. 437).

Trimble, S. W. 1999. Decreased Rates of Alluvial Sediment Storage in the Coon Creek Basin, Wisconsin 1975–93. *Science* 285:1244–1246.

NOTES

1. A spatial family as used here is not to be confused with a taxonomic family used in soil taxonomy or other soil classification systems.
2. Only well-drained and poorly drained conditions are identified in Figure 7–12. More detailed soil classifications identify "excessively well-drained" soils for sandy soils with deep water tables, "moderately well-drained" and "somewhat poorly drained" as intermediate classes between well-drained and poorly drained classes, and a "very poorly drained" class for soils that have a water table at or very near the surface.

CHAPTER REVIEW PROBLEMS

1. Sketch the cross section of a landscape, label the various parts of that landscape, and identify which part(s) are most likely to erode or flood.

2. What type of land is in danger of severe erosion? What can be done to protect land from eroding?

3. How does a riparian buffer protect surface waters from sediment and nitrate?

4. What physical principles of erosion control are practiced in no-till management?

CHAPTER

8

Activities on the Land

Land and soil have always been intimate components of human activity. Almost all early writings about soil relate the influence of water, organic residue, and manure to the growth of crop plants.[1] Many human activities that change what we see on the land are easily recognized. Hills are leveled to facilitate roads and subdivisions, areas are flooded as dams are constructed to create lakes, and forests are cut to make way for numerous other uses of the land. Land also changes over time in response to natural events. Natural changes to land occur both as sporadic catastrophic events so rare they escape human observation or even recorded human history and at rates so slow that the changes are imperceptible in the relatively short time span of human life.

Most land is covered with vegetation. We see approximately half of all plants. The roots that support the aboveground portions of the plant are seldom seen. We see buildings but seldom their foundations, although we realize that a foundation anchored in the soil is present. A building may collapse because of a structural failure in the foundation. We can see the larger mineral particles in the soil, but we do not see the chemistry or the microbial life present within the soil. We visually recognize the role of chemistry and microbiological activity in soils only through the influence they have on growing plants. If plants fail to grow as expected, it often is because of unseen chemistry or microbiological activity in the soil. Humans in response to ambient chemical, physical, and biological conditions of the land and soil in various parts of the world have adopted a multitude of customs and practices that enable them to sustain life. Most of the customs and practices that humans employed to obtain food and other necessities have evolved by trial and error within localized areas of land and temporal social and economic situations.

When people view unfamiliar land, they often find it difficult immediately to understand the reasons for what they see people doing on the land. When people from temperate latitudes see people living in what appear to be flimsy thatch-covered houses in the tropics, they may have concern for the well-being of the residents during the cold winter. Their concerns are unfounded because temperatures vary little throughout the year in the tropics. Conversely, when people in the tropics see snow-covered fields they may have concern for how the people living there are able to grow food. They often consider the construction of brick houses as extravagant, not realizing the rigors of winter in temperate latitudes. Just as people adopt the design of houses to accommodate the extremes of climatic conditions, biological ecosystems have developed to accommodate the extremes of environment experienced at any given site. We often conceive of soil and land as something permanent but fail to recognize the dynamics of soil and land that determine biological ecosystems and human use of the land.

Human activity on the land results from a combination of the properties of the land and soil and the ability of people to alter those properties in ways that are compatible with human needs and desires. Human ability to alter land and soil or adapt to different land and soil has evolved throughout history and differs greatly throughout the world at the present time. Many of these differences can best be attributed to social, political, and economic conditions that are discussed in greater detail in Chapter 9. In this chapter we examine a few of the long-term natural changes that shape the land and the basic challenges land and soil properties present for humans and how those challenges are met.

HUMAN REQUIREMENTS FROM THE LAND

Humans have three basic requirements from the land they inhabit: temperature, water, and chemical elements necessary for their physiological functions and physical growth. Land and soil have acquired their properties by being anchored in one place for long periods of time. Dynamic events that have sporadically occurred during that time have determined many of their present properties and have a major influence on both human and natural ecological habitation of the land. Associations are often made between average land and soil conditions and life-forms. Averages provide us with correlations between human activity and land properties, but extreme events are often more determinant of human activity than averages.

Many characteristics of land can be traced to the influence of climate. Climate is the sum total or average of all weather conditions usually presented as yearly or monthly averages. Do averages really do an adequate job of identifying the environment within which soil and life-forms exist? Before we respond, perhaps we should consider how an average might affect us personally. Most of us feel comfortable with bath water of about 122°F (50°C). But were we to have one foot in a tub of 32°F (0°C) iced water and one foot in a tub of 212°F (100°C) boiling water, comfortable would not describe our condition, even though 122°F (50°C) is the average temperature of our bath waters. In similar fashion, it is not the climatic or average weather conditions that determine the features of a soil or the vegetation growing on the land but rather the extreme weather events. As in the case of our bath water where one foot would numb from the cold and the other blister from the boiling water, humans and other life-forms must be able to withstand the extremes of weather conditions to survive and flourish at a given site.

Water is necessary for all forms of life. Average annual temperatures and average annual amounts of rainfall are but crude guides to vegetative production. The successful cultivation of food crops requires only that desirable temperatures and adequate amounts of water suitable for crop growth coincide for a period of time necessary for a given food crop if farming is to be successful (see Chapter 3). A reliable supply of available water during the times when temperatures are suitable for plant growth is more critical than total annual rainfall. The frequency at which water is available during the growing season is also critical. If gentle rain fell each day in proportion to the rate at which water evaporated and transpired from the growing plants, the soil would need to be only a few inches deep to accommodate the plant roots and surround them with enough water to satisfy growing plants. If, instead of a daily gentle rain, the same amount of water fell during only one day each month throughout a year, an entirely different chain of events would take place. Plant roots would have to access a depth of soil sufficient to acquire an adequate supply of water for the some 30 days between rainfall events. Most climatic records, however, would record the same average monthly and annual rainfalls for both scenarios.

The conditions just described are extreme and not entirely realistic. Most climates are intermediate between such extremes, but many soil characteristics and human attempts to grow food are determined by the extremes in weather events, and precise correlation between soil properties and average weather (i.e., climatic) conditions is problematic. In the scenarios described it is probable that if all the annual rainfall occurred during one day each month, much of the water would run off even the gentlest slope. If all of the once-a-month rainwater were able to infiltrate, it may wet the soil to a great depth removing or leaching soluble chemical components to depths below the reach of growing vegetation roots. If frequent but small amounts of rainfall daily wetted the soil but was evaporated or transpired through the leaves of the vegetation, soluble salts from dissolving soil minerals not ingested by the plant roots would accumulate around the root. This would increase the osmotic value of the soil solution, making water uptake by the plant increasingly more difficult, and salinity would eventually kill the plant.

GEOLOGIC BLOODLINE OF SOIL

The effects of water and temperature on plant growth are fairly easy for most people to see and at least partially understand. It is more difficult to understand the need for the complete spectrum of elements necessary for both plant and human physiology and growth. As discussed in Chapter 5, many of these elements are available only from the dissolution or weathering of minerals in the soil. No soil can escape the composition of its mineral pool imparted to it by the geologic material from which it formed. Certain chemical characteristics of the mineral pool can be modified, but certain inherited characteristics remain until a soil is either dug up and removed or eroded.

There can be little doubt that early humans did not have knowledge of the role of essential elements in the production of food crops. They must have relied on trial and error as they migrated in search of land that could sustain their efforts to grow food crops. Among the lands that had the most readily observable necessities of temperature and water, they encountered an array of soils with various mineral compositions related to geologic formations. Among the many geologic materials from which soils may form, a few stand out as distinctive bloodlines, so to speak, among all soils. Deposits of sand, such as beach dunes or point bars in meandering stream channels, dictate that the soils formed in them will be of sandy texture. If that sand is primarily quartz (SiO_2), which it most often is, no amount of weathering can create essential nutrients such as phosphorus, potassium, calcium, or magnesium needed for plant growth. Soils formed in only quartz sand will, in addition to being drought susceptible because of their inability to retain plant-available water, be chemically infertile and not able to sustain the growth and harvest of human food crops.

In a similar but contrasting scenario, clayey parent material, like former lake beds and deltaic deposits, can never be sandy-textured soils. The chemistry of soils formed in clayey sediments may differ depending on the composition of the clay and the chemistry of the water infiltrated into the soil, but many contain adequate quantities of essential elements to sustain the harvest of several food crops. The origin of many clay-textured materials is the deposition of eroded soil material as streams and rivers enter lakes, deltas, and floodplains. Clay deposits often contain organic residues eroded from the surface of soils in the watershed. As the organic residues decompose, they provide nitrogen and other plant-essential elements and are often desirable sites for food crop growth.

Distinct chemical properties are present in soils formed from limestone. Limestone varies in composition, but calcite ($CaCO_3$) or dolomite $Ca:Mg(CO_3)_2$ minerals are principal components. If the limestone is nearly pure calcite, the soils formed will be very thin because exposure to carbon dioxide and water causes the calcite to dissolve, forming calcium bicarbonate that is soluble [$CaCO_3 + CO_2 + H_2O \leftrightarrow Ca(HCO_3)_2$] and is removed with water leaving the area either as surface or subsurface flow. As groundwater flows through limestone, underground caves are often formed (Figures 8–1). Limestone is often rich in phosphorus because it contains the skeletons of marine life so concentrated phosphorous is mixed in the sediment on the organisms' death. Most limestone contains admixtures of silicate clays, silts, and sands cemented by calcite or dolomite. In such limestone, sometimes called dirty limestone, the included silicate materials are left as the calcite dissolves, and deep soils composed of silicate materials are often formed on level land or accumulated on the toe-slope and bottom-land positions that receive sediment and run-on water from adjacent eroding limestone hills. With few exceptions, soils formed from the various types of limestone are chemically rich in calcium, magnesium, and

FIGURE 8–1 Photo of limestone hills in the People's Republic of China. Note the cave entrance formed as underground water dissolved the limestone.

FIGURE 8–2   **Photo of plowing near limestone hills (*karst*) in the People's Republic of China. It was reported that these fields have been continuously farmed for many centuries.**

phosphorus, and have pH values near 7, thus assuring maximum availability of these mineral-derived essential elements (Figure 8–2).

The thin, rocky, and seasonally dry soils of Greece and other areas surrounding the Mediterranean Sea are striking examples of soils formed from limestone and marble, a metamorphosed or hardened form of limestone, that have nourished the food crops for some of the most culturally advanced ancient civilizations. Similar associations can be made between limestone-derived soils and the Inca and Mayan civilizations in South and Central America and ancient cultures in China. The natural fertility of soils in the Midwest of the United States and Eastern Europe are in large part formed from limestone crushed and deposited by continental glaciers.

In stark contrast with soils formed from limestone are soils formed from igneous rocks. Igneous rocks include granite, diorite, rhyolite, andesite, and other rocks formed directly from the cooling of the molten magma of the earth's core. Metamorphic rocks such as schist and gneiss are formed from compression of primary igneous rocks and are of similar chemical composition and differ mainly in mineral type and structure. Igneous and metamorphic rocks have not been enriched with life-essential elements. The chemical composition of igneous rocks differs, but compared to limestone, most contain relatively small quantities of many of the elements necessary for biological life. Soils formed from igneous materials therefore inherit lesser concentrations of the essential elements required by growing plants and animals than sedimentary rocks such as limestone.

FIGURE 8–3 Photo of a large buttressed tree in the Amazon jungle in Peru, SA.

Where climatic conditions are compatible with plant growth, large forests of slow-growing trees often became established on soils formed from igneous rock materials (Figure 8–3). Humans frequently interpret the presence of massive trees in the forest vegetation as indicating a fertile soil. Most massive trees have rather slow growth rates and require only a slow release of essential elements from the soil minerals. They tend to be more prevalent in relatively infertile sites, quite possibly because more rapidly growing species did not flourish with the limited nutrient supply, thereby allowing the giant trees to attain dominance and shade smaller species. When humans attempt to grow and harvest food crops from these soils, they quickly become chemically impoverished because the limited store of essential elements in the organic pool is harvested in the food crops. Historically, human civilizations were sparse in areas where soils are formed from igneous rocks and human population density remains low except where food crop production is maintained with fertilization or food is imported.

If soil properties could be directly related to the composition of the geologic material from which they form, there would be no need for soil science. We would need only to study geologic materials. However, soils are located at the surface of the land above the geologic rocks. In that position the minerals of geologic origin contained in soil are altered by interactions with the environment of the atmosphere above and life-forms of organisms that live in or that take root in the soil. We must also recognize that environment near the land surface physically moves soil material on the land.

NATURAL DYNAMICS OF LAND

Over long periods of time, there is natural removal of soil material and formation of new soil material on most land. As soil material is removed, a new volume of material is exposed from which a new generation of soil is formed. The physical removal of soil from one location often results in the deposition of soil material at another site on the land where a new soil, different from the soil that eroded and different from the one buried by the deposition, will form within the deposited material. This natural redistribution of mineral material by erosion and deposition has given rise to a diversity of soils within rather small land areas throughout the world. The physical movement of soil can be either dramatic, as in landslides (Figure 8–4), imperceptibly slow by gradual downhill movements called *creep*, or the spasmodic erosion of material by wind and/or water. Depending on the rate at which soil material moves within the landscape, the life span of an individual soil at a specific location on the land may be from less than hundreds of years to many thousands of years, creating a false concept of permanence to human understanding.

FIGURE 8–4 Photo of a landslide area in undisturbed woodland in western North Carolina, USA. The landslide area is approximately 100 feet (30 meters) wide and extends for approximately 800 yards (731 meters) straight down the mountainside. A total of about 50 such slides were observed in a county of approximately 200,000 acres (81,000 ha) following two days of excessive rain.

The chemical dissolution of minerals is a largely unseen mechanism of removing soil material. As water leaches through soil or flows over the soil surface, small amounts of the minerals in the soil dissolve and move with the water. Soil loss by dissolution is almost totally undetected on the land but can be measured as dissolved elemental salts in river water. Dissolved salts are concentrated as water evaporates in oceans and inland seas such as the Dead Sea on the boundary between Israel and Jordan and the Salton Sea in California, USA. Physical movement of soil material by erosion is much more visible.

Judson and Ritter (1964) provide an overview of both physical and chemical removal rates within the major watershed areas in the United States (Table 8–1). By analyzing both dissolved elements and suspended solid materials contained in river water, flow rates, and the total land area drained by several rivers, they calculated the rate at which land surfaces would be lowered throughout the various river basins by erosion of solid soil particles and leaching of dissolved elements. Their calculations of soil loss were made from data acquired during various lengths of time prior to 1960 and assume that erosion and dissolution were uniform on all land within each watershed.

Note that solid removal (erosion) rates are greatest in the mountainous, mostly arid, and sparsely vegetated Colorado River watershed with little cultivated area. Lower rates of erosion were found in the more level and more humid watersheds of the Columbia, Mississippi, South Atlantic, and eastern Gulf of Mexico watershed areas where much of the land was cultivated. It should also be noted that in the Columbia River watershed and in the watersheds in the humid southeastern United States, more material is lost by dissolution and leaching than by the erosion of solid particles.

TABLE 8–1	Dissolved and Solid Removal Rates Within Major Watersheds in the United States			
Drainage Region	**Dissolved**	**Solid**	**Total**	**Total Rate of Removal**[*]
	$T\ mi^{-2}\ Yr^{-1}$	$T\ mi^{-2}\ Yr^{-1}$	$T\ mi^{-2}\ Yr^{-1}$	*In 1,000* Yr^{-1}
Colorado River	65	1190	1255	6.5
Pacific slopes (California)	103	597	700	3.6
Western Gulf (Texas)	118	288	406	2.1
Mississippi	110	268	378	2.0
North Atlantic	163	198	361	1.9
South Atlantic and eastern Gulf	175	139	314	1.6
Columbia River	165	125	288	1.5
Total USA	121	340	461	2.4

[*]Calculated as uniform removal from the surface of all the land in the river watershed. T = tons; mi = miles.

Source: Judson and Ritter, 1964.

TABLE 8–2   Erosion Rates of Solid Particles within Individual River Basins

River Basin	Location Measured	Erosion Rate (In $1,000 Yrs^{-1}$)*
San Joaquin	Vernalis, CA	0.1
Tombigbee	Jackson, AL	0.5
Alabama	Claiborne, AL	0.5
Sacramento	Sacramento, CA	0.5
Columbia	Pasco, WA	0.5
Rappahannock	Remington, VA	0.7
Snake	Central Ferry, WA	0.7
Delaware	Trenton, NJ	0.8
Mississippi	Baton Rouge, LA	1.3
Rio Grande	San Acacia, NM	1.8
Pecos	Puerto de Luna, NM	3.6
Colorado	Grand Canyon, AZ	5.6
Mad	Arcata, CA	19.3
Eel	Scotia, CA	30.4

*Calculated as uniform removal from the surface of all land in the river watershed.

Source: Judson and Ritter, 1964.

A more detailed analysis of erosion rates within the watersheds of smaller rivers reveals considerable variation within each region (Table 8–2). It is clear that the erosion rates as measured in the sediment load in rivers are greater in more arid and mountainous regions than in more humid and less mountainous watersheds. In most cases sparse vegetation associated with aridity appears to contribute to increased erosion. The surprisingly high rates of erosion within mainly forested watersheds of the Eel and Mad Rivers in Northern California result from steep slopes and tectonic instability. Although containing some steep slopes in the upper reaches of the watershed, the central portion of the San Joaquin river basin in California is a nearly level floodplain. Although the basin is intensely cultivated, little erosion is recorded from the watershed. As discussed in Chapter 7 with the data from the Coon Creek watershed, erosion rate on the upland and sediment load in the rivers cannot be reliably linked because of intervening deposition of eroded material on floodplains. In very steep land, rivers have steep gradients, rapid flow rates, and narrow floodplains, so essentially all eroded material is suspended as solid particles and exits the area in the river water.

The measurements of both solution and solid loss of soil material via river flow point out that all upland land surfaces are slowly being lowered. Although the rate of lowering may be imperceptible in human experience, lowering of the land surface is a natural process that allows plant roots to explore and seek essential nutrients from a new volume of soil minerals.

Long term, erosion and dissolution rates averaged over entire watersheds may be viewed much in the same manner as average rainfall and temperature data. Within each watershed some areas are subjected to rapid rates of erosion while others are only slightly affected. Other land, usually adjacent to the eroding site, receives the eroded material in the form of sediment. Each new layer of sediment is usually quite thin and quickly acquires the characteristics of the former soil surface. Most severe erosion and deposition events occur infrequently during what are often termed *one in twenty year* or *one in a hundred year* weather events. Although these time spans are long in the human time scale, they are short in the time scale of soils and landscapes where major tectonic alterations of landscapes may take place only once in many thousands of years. Even more drastic events take place during longer spans of time as witnessed by the exposure of at least four buried soils in southern Arizona (Figure 8–5). Each buried soil pictured has characteristics indicating that the land surface was stable for a long enough period of time for vegetation to grow, organic matter to accumulate near the soil surface, and weathering to release iron and form reddish colored subsoil before burial. It is not known how much time elapsed during which each soil formed, was successively buried, and then exposed to reveal their presence, but a preliminary estimate of at least 30 thousand to perhaps 100 thousand years was made on the basis of the material exposed at the base of the *arroyo*.

FIGURE 8–5 Photo of a recently eroded *arroyo* bank in southern Arizona, USA, exposing four distinct soils (buried soil surfaces indicated by arrows) that formed on former land surfaces and were subsequently buried.

Humans are acclimated to the much more rapid life cycles of animals and plants and have difficulty conceptualizing the life cycles of soils. People see only short time frames in the processes of soil and landscape formation. People tend to conceive of erosion or depositional events as catastrophic rather than natural processes in the continuum of the much longer life cycle of soil. Soil will continue to be present on the land so long as there is a volume of mineral material capable of supporting plants at the earth's land surface. Each event in the life cycle of soil produces a new soil that is somewhat different from surrounding soil and contributes to spatial diversity in the family of soils on the land.

The life cycle of soil, although long by human standards, is brief within geologic time. Material dissolved or eroded from soil may stop moving and become a component of another soil several times before it eventually finds its way to the ocean. In a very long-term or geologic view of time, the nutrient elements contained in the suspended or dissolved river waters that enter the ocean are not lost but on the way in a long-term geologic cycle of elements on planet Earth. In the ocean, life-essential elements provide nourishment to a food chain of aquatic organisms that on their death fall to the bottom and become part of limestone, shale, or sandstone. As the tectonic plates of the earth's surface shift, ocean sediments are thrust above the sea in the form of limestone, sandstone, or shale material to take their place as a volume of the land surface to again be converted into soil and undergo another soil life cycle. Approximately 66% of the present land surface consists of former sea bottom sediments now present as limestone, shale, and sandstone. Although people may think of land preservation in terms of decades or even centuries, within geologic time the land is constantly changing. Only the elements are preserved as they ride on the merry-go-round of natural processes through nearly infinite millennia.

HUMAN ALTERATIONS TO THE LAND

Humans have two choices in seeking the food they need for life. Either they can roam among the natural vegetative ecosystems and gather food or they can accumulate the food plants and animals they desire and culture them. Clearly, the course of human society has been toward domesticated and cultured food supplies. To acquire land for the domestication of food crops, natural vegetation and animals had to be removed. It is probable that humans have extensively altered natural ecosystems on the land since they learned to create fire. The shade of natural forests had to be removed to provide the sunlight necessary for food crops and pasture grasses for domesticated animals. Also aggressive species of wild animals needed to be controlled to protect human lives as well as those of domesticated cattle (Figure 8–6).

To make land more uniform for the growth of their food crops, humans first remove existing vegetation. The surface of the soil must be disturbed so that seed can be planted. Early settlers that cleared the forests in the United States often used the expression "tame the soil" to express the homogenizing of spatial microvariability found, and often created, when trees were dug out to make way for cultivation. The first crops planted on so-called untamed land often had very uneven growth until a few years of cultivation mixed the top layer of soil and made it more uniform. The problems faced in newly cleared land are not unlike those faced by the owner of a

FIGURE 8–6 Photo of a lion in Kenya. He cannot be a welcome visitor in your garden or cornfield.

newly constructed home where the builder indiscriminately bulldozed trees and distributed clayey subsoil and rock removed from the house foundation onto the site planned for the lawn. The homeowner or builder often solves these problems by importing and spreading a layer of topsoil to provide a more uniform soil for the lawn.

Termite mounds are a very visible example of spatial heterogeneity in some tropical ecosystems. Colonies of termites gather soil material from surrounding areas and construct mounds to protect themselves. In the process they create concentrations of soil material and life-essential elements. The presence of termite mounds (Figures 8–7 and 8–8) in some tropical areas bedevils planters as they attempt to obtain uniform growth in newly cleared fields. The mounds are not only a physical impediment to domestic cultivation but also present contrasting areas of fertility when destroyed in cultivation efforts.

HUMAN INTERVENTION TO OBTAIN WATER

To secure a reliable supply of water for cultivated food crops, humans attempt to control water rather than rely on dynamic and unpredictable amounts provided by rainfall. The construction of massive dams to store surface water in artificial lakes for irrigation is well known. Humans also seek to obtain more land suitable for their food crops by protecting land from water by building levees and dikes such as the famous polders in The Netherlands.

FIGURE 8-7   Photo of a termite mound in Ghana.

Providing water for plant growth is one of the oldest human innovations. Extreme dryness dictates that water necessary for the production of food crops must be obtained from a source external from the crop field. Many forms of irrigation are practiced throughout the world. The most easily accomplished form of irrigation is the diversion of surface waters from rivers via a system of constructed canals so water can be distributed over a larger area of land. This was most easily done on nearly level floodplains adjacent to major rivers. In some parts of the world, elaborated canal systems were constructed in mountainous areas to collect runoff water from steep slopes and channel it to more level land where the water could be evenly spread over large areas devoted to growing food crops.

Where there is a paucity of nearly level floodplains, steep slopes are often terraced to provide level areas that retain water that would otherwise run off during storm events, as discussed in Chapter 7. Often these bench terraces are connected to canal systems that collected water from higher on the mountain slopes to provide irrigation during rainless periods.

On level land where the water table was near the surface, human labor and simple machines lifted water from dug wells and distributed it to crops on the land surface

FIGURE 8–8   Photo of large termite mounds in Zambia.

(Figure 6–3). As more sophisticated pumps and powerful electric and fuel-powered motors became available, it became possible to access very deep groundwater in some locations. Irrigation systems that travel over the land have permitted the application of water to fields that are not level enough to distribute water by *flood irrigation*. This eliminates the need to level and smooth the land surface. Many of these irrigation systems, often known as *center-pivot* systems, travel in a circle around the well from which the water is pumped, as seen in Figure 8–9. The depth to which groundwater can be pumped for irrigation via modern machines becomes an economic balance of pumping cost versus the value of the crop grown. In many areas groundwater has been pumped more rapidly than it is restored, and irrigation systems have been abandoned as pumping lowered the groundwater table to depths where pumping costs became excessive.

Too much water in the soil, that is, a water table near the soil surface, presents several challenges to human food production in many areas of the world. Again, humans have been able to understand and to some extent control excessive wetness in many areas. Excess wetness most often is present in broad level areas of land and can be removed by construction of artificial channels or canals to lower the water table. If there is a place at a lower elevation to deposit the excess water, drainage canals can operate by gravity flow. In some areas powerful water pumps pump excess water some distance to rivers, lakes, or oceans.

Extensive systems of drainage canals have been constructed in many parts of the world. Some of the most notable are in Western Europe, Asia, and swamp areas of the

FIGURE 8–9   Aerial photo of circular *center-pivot* irrigation systems in western Texas, USA.

Atlantic and Gulf coastal plains of the United States. Controlling the depth of the water table in excessively wet areas of level land has provided some of the most productive land for crop production in the world. Rapid removal of excess water in the soil allows air to enter the large soil pores within the root zone and eliminate crop damage from saturation. Constructed drainage systems also provide a mechanism to protect crops from short periods of drought. Over half of the cropland in many parts of the United States uses some form of human-constructed drainage (see Chapter 6, "Too Much Water: Hydric Soils").

HUMAN QUEST FOR TEMPERATURE

Where temperatures are too cold for growing food plants, people are forced to rely on food sources coming to them in the form of fish or animals that can graze on the existing plants and convert those plants into meat for human consumption. Only recently in human history have people been able to provide heat, at great expense via artificial temperature control in heated greenhouses in cold areas. High temperatures are seldom a problem in crop production, except for brief periods of time during the growth of some crops. In temperate latitudes, people can escape temperatures that are too high for desirable food crops by timing the planting of food crops with the dynamics of seasonal temperature changes. The human response to temperature limitations is to avoid those lands not compatible with food production or rely on external sources of food.

HUMAN QUEST FOR ESSENTIAL ELEMENTS

Humans have a consistent tendency to remove the most nutrient-rich portion of the food plants they grow, and the essential elements contained are discarded as waste near home sites and cities and not returned to the site of crop production. Removal of essential elements in the harvested plants interrupts natural biocycling, and therefore the return of essential elements to the land where the crop plants were grown. Transporting human food from the site where it is grown rapidly depletes the essential elements contained in the available and organic pools in the soil (Figure 5–1; Table 5–3).

The soil responds to this human harvest of essential elements by supporting less vigorous plant growth. The surface of the soil is thus more exposed, erosion rates increase, and as surface material is removed, minerals deeper in the soil are exposed to weathering and exploitation by plant roots. If the geologic materials exposed by erosion weather rapidly, as in nutrient-rich limestone, limestone-rich glacial deposits, and certain other materials, the soils remain relatively fertile despite elemental depletion by harvest and erosion, and humans are able to obtain crop yields sufficient for continued habitation. Parts of central China and Eastern Europe are examples of this scenario. Where the mineral pool in the soil is relatively poor in essential elements or the rate at which nutrient-bearing minerals weather is too slow for sustained food crop production, humans are forced to abandon the land after only a few crops are harvested.

It is possible to replenish the elements harvested by returning organic waste material to the fields. Even soil material that is naturally infertile can be enriched by the importation of essential elements contained in organic waste materials grown on other sites. Examples of this can be found around human settlements where hunting and fishing provide a large portion of human nourishment and the waste materials are discarded on the land near the settlements, creating fertile sites for food plant growth.

The practice of using domesticated animals, usually goats, sheep, or cattle, to range over extensive areas of land and graze plants unpalatable for humans and daily herding those animals into confinements near villages permits the gathering of manure. The manure is then applied to small areas of land, chemically enriching that land for human food crop production. Even the most naturally infertile soils can grow more luxurious vegetation than naturally possible when humans are able to enrich them with the essential elements needed by plants.

All practices that chemically enrich naturally infertile soil by importing essential nutrients from adjacent land have spatial limitations determined by the ability to gather and transport the essential elements to sites of food plant growth. The need for rather large areas of land from which natural vegetation can be harvested and concentrated limits the density of human habitation on soils formed from chemically poor minerals. Examples of human activities that enrich small areas of relatively infertile soil are found in the warm, moist tropical Amazon jungle around now extinct indigenous villages. These areas, known as Amazon Dark Earth (Terra Preta) are usually located on nonflooding land near rivers. The indigenous people augmented their diet with fish (Figure 8–10) and game and distributed their wastes around the village, creating soils that contain more essential elements than surrounding soils and black color from the charcoal remains of their cooking fires (Lehmann et al., 2003).

FIGURE 8–10   Photo of a large catfish lying in a natives' dugout canoe on the bank of the Napo River in eastern Peru, SA. Note the pocketknife inserted for scale reference in the wound inflicted by the native fisherman.

SPECTRUM OF HUMAN CHALLENGES ON THE LAND

The challenges to human food production on the land surfaces of the earth can be summarized as temperature, water, and essential elements. Although humans are able to survive physically for short periods of time in the most severe temperature extremes, only sparse populations have flourished in the extreme coldness of polar latitudes and highest mountains. Without the ability to grow plants for human food, their diet depends on availability of migrating animals and fish or on sophisticated infrastructure for importing food. Challenges posed by too much and too little water have been met by drainage of wetland and irrigation of dry land, respectively. Technologies that improve the efficiency of both irrigation and drainage have improved over the centuries.

Chemical challenges to human food production are responsible for much of the human habitation patterns in areas of the world where temperature and moisture are compatible with human food production. Unaided by chemical technology and an understanding of chemistry needed for growing plants, indigenous peoples are forced to use trial-and-error methods to find chemical soil fertility. As humans sought to abandon hunting and gathering lifestyles and grow their food, they moved from place to place until they were successful in growing desired food crops or sustaining herds of domestic animals. When a location was found that permitted them to grow food crops reliably, they established villages and planted their crops in surrounding lands.

WARM, WATERED, AND EXTREMELY INFERTILE LAND

A broad continuum of chemical fertility naturally occurs among the spectrum of soils with adequate water and temperature for crop growth. Some areas are so limited in chemical fertility that only very sparse human habitation is possible. Perhaps the most extensive example of such land is approximately 400 million acres (200 million ha) in central Brazil south of the Amazon jungle. Although rainfall and temperature are adequate for forest vegetation, as evidenced on rare outcrops of limestone or basalt in the area, the dominant native vegetation is a sparse growth of small trees and grasses known as *cerrado* (Figure 8–11).

Most soils in these areas are Oxisols and the poorest of Ultisols (see the appendix). These soils are formed from the weathered remains of acid igneous (felsic) rocks like granite that have been exposed at the surface of the land for perhaps billions of years. During that time the minerals have been subjected to decomposition during repeated cycles of erosion and deposition. Almost all of the minerals that originally contained essential elements such as calcium, magnesium, and potassium have been weathered and removed. The present soils have formed in thick layers of material composed almost entirely of iron and aluminum oxides, kaolinite clay and quartz sand (Figure 8–12). Phosphorous contents are naturally low in igneous rock, and most phosphorus in such soils is fixed as iron and aluminum oxide minerals from which the release of nutrients to the available pool in the soil is too slow for the growth of human food plants. Natural vegetative growing in these areas is limited by the lack of nutrients, and therefore the organic residues returned to the soil contain low amounts of essential elements.

Although abundant grass is often present among the small trees, the natural vegetation is so poor in calcium and phosphorus that it supports no indigenous herds of large

FIGURE 8–11 Photo of natural *cerrado* vegetation in Brazil, SA.

FIGURE 8–12 Photo of a road bank in central Brazil, SA, revealing more than 12 feet (4 meters) of nutrient-poor sediments, with an Oxisol soil formed at the surface. The site is typical of the chemically inert materials of quartz sand, iron oxide, and kaolinite clay in which most of the soils in the *cerrado* have formed.

animals so common on the savannas with similar climatic conditions in Africa where soils are formed in nutrient-rich geologic materials. Most attempts to graze domestic cattle on *cerrado* vegetation failed. The area became known as a place to lose cattle rather than a place to graze cattle. The native grasses that were able to grow on extremely low contents of available calcium and phosphate in the soils would not sufficiently nourish cattle. A common cause for cattle loss in the *cerrado* was broken bones as they grazed on the calcium- and phosphorus-deficient grasses. After only a few months of grazing on *cerrado* vegetation, the cattle's bones became weak and there was severe mortality from broken bones. Indigenous ranchers often burned mature grass to quickly recycle essential nutrients into palatable new grass with only marginal success as seasonal rainfall patterns precluded new growth for several months each year. The calcium and phosphate deficiency in the native grass could be overcome only by importing concentrated mineral supplements as part of the animals' diet.

Attempts to grow food crops in the *cerrado* completely failed. Only sparse human populations, primarily miners, lived in the area. Extensive experiments were conducted in the 1950s using amounts of fertilizer deemed sufficient and economically profitable for potential farmers. The amount of fertilizer used in the experiments was determined from successful corn, soybean, and other grain crop production on more naturally fertile soils in other parts of the world. Essentially all of the experiments failed to produce adequate crop growth. In 1964 there was almost no intensive farming in the *cerrado*, and the conclusion was advanced that the land was unsuitable for agricultural crop production (Wright and Bennema, 1965). Although the early experiments applied amounts of phosphate fertilizer adequate for most growing crops on most soils, the fertilizer phosphorous applied was quickly fixed from the available pool into unavailable forms in the mineral pool by the presence of high quantities of iron and aluminum oxides present in most soils in the *cerrado* area.

In the 1970s, a new group of soil scientists again attempted fertilizer experiments with corn and soybeans in the *cerrado*. Armed with chemical and mineralogical measurements that determined the potential these soils had to fix soluble fertilizer phosphorous, they applied more than ten times the rates previously used. When these high rates were applied in conjunction with adequate supplies of the other essential elements and lime to adjust the naturally acid condition of the soil, the results were dramatic (Figures 8–13 and 8–14).

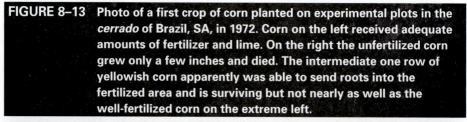

FIGURE 8–13   Photo of a first crop of corn planted on experimental plots in the *cerrado* of Brazil, SA, in 1972. Corn on the left received adequate amounts of fertilizer and lime. On the right the unfertilized corn grew only a few inches and died. The intermediate one row of yellowish corn apparently was able to send roots into the fertilized area and is surviving but not nearly as well as the well-fertilized corn on the extreme left.

FIGURE 8–14   **Photo of the first crop of soybeans on experimental plots growing on extremely infertile Oxisols in the *cerrado* of Brazil, SA, in 1974. Note the soybeans in the foreground growing on unfertilized soil are only a few inches high and will not survive, whereas those more distant from the camera, growing where adequate lime and phosphorus fertilizer were added, are healthy and on their way to producing a good yield. Natural *cerrado* vegetation is seen in the background.**

Phosphorus fertilizer is expensive. The cost for the necessary phosphorous, lime, and other fertilizers per acre was approximately four times the cost of buying an acre of land.[2] Continued experimentation determined that only one rather massive amount of phosphorous was able to correct the capacity of the soil chemically to fix phosphorous. After the initial application it was only necessary to apply annual rates phosphorus no greater than the amount harvested in the crop. Thus the large initial investment in phosphorus fertilizer was a capital investment in the land, not an annual expense. A few very competent farmers from southern Brazil seized on these results, and by 1985 a few large privately owned fields began to appear in the *cerrado* (Figure 8–15). As bankers and other investors came to understand that phosphorus fertilizer when used to correct the soil's natural infertility was not an annual fertilizer expense in the usual sense but rather a onetime capital investment that made a permanent improvement in the land, dramatic changes came to the area. Initial investments in fertilizer could be considered like other capital investments such as irrigation systems or drainage systems rather than annual costs, and money was made available to farmers interested in venturing into the *cerrado*. Note that because there were no indigenous farmers in the *cerrado*, almost all those who started farms were

FIGURE 8–15   Photo of nearly mature soybeans in a farmer's field that received adequate lime and fertilizer in the *cerrado* of Brazil, SA, in 1986.

experienced farmers or often the sons or daughters of experienced farmers from other parts of Brazil knowable in the business aspects of commercial farming. By the mid-1990s, with approximately 20 million acres (10 million ha) of the *cerrado* under cultivation, the area was producing 40.5% of the beef cattle, 43% of the soybeans, and 28% of all grain production in Brazil (Lopes, 1996). In 2003 Brazil became the world's largest producer of soybeans, surpassing soybean production in the United States (Haverstock, 2004).

Like every other land area, the soils in the *cerrado* are not all the same. There are a few areas where the soils are formed from basalt or limestone. These areas were originally covered with seasonal rainforest vegetation of large trees. Early European settlers attempting to find farmland in Brazil found that they could have success if they cut and burned these forested portions of the *cerrado* area as opposed to complete failure in the true *cerrado* vegetation. After approximately 100 years of European settlement, most evidence of forest vegetation had been removed, but occasionally a fairly successful looking farm or area of farms would be seen in the *cerrado*. The one feature that appeared common to most such areas was the presence of a few palm trees (Figure 8–16). In 1968 the author was told that it was the indigenous practice in the *cerrado* to retain a few palm trees when clearing a forest because they indicated fertile soil and their presence would allow the owner to command a higher price were the land to be sold. During the past 35 years, fields with palm trees have almost entirely disappeared in the *cerrado* region perhaps because they got in the way of machinery or just died of old age, and reliable soil chemical analyses have replaced them as indicators of fertile soil.

FIGURE 8–16   Photo of natural palm trees preserved in a cultivated field as indicators of the former forest vegetation and good natural fertility in the *cerrado* region of Brazil, SA, in 1968.

WARM, WATERED, AND SLIGHTLY FERTILE

Soils that are only slightly more fertile than those in the *cerrado* often naturally support rain forest vegetation in the tropics and woodlands in temperate latitudes. Most of these soils are also acid in reaction. Big slow-growing trees with extensive root systems tend to dominate the naturally occurring mature forests. The presence of big trees is often mistaken as evidence of a fertile site, but in reality the soils are usually nutrient poor. Several reasons have been advanced for the presence of large trees on relatively infertile soils. It is probable that large tree species with extensive root systems can obtain nutrients from a large volume of soil, and their ability to extend a canopy above other species and thus shade the smaller species have advantages in such soils.

When forests are cut and the dried biomass burned, the fire releases nutrients from the organic biomass into the available pool and the site is immediately able to support vigorous plant growth. To avail themselves of this temporary increase in soil fertility, people needing food crop production routinely burn many old-growth forests. This land use practice, often called *slash and burn* or *shifting cultivation,* has been and is extensively use in many of forested areas to support very limited human populations both in the tropics and in warmer areas of temperate latitudes primarily in southeastern areas of Asia and the United States. Where human population density is low and large amounts of land are available, this system is sustainable.

When experienced slash-and-burn farmers are not confined by property boundaries, they select a fertile site by estimating the nutrient content of the biomass, or more importantly the nutrient content of the ash that will remain on the soil when the biomass is burned (see Chapter 5, "Essential Elemental Flow and Indigenous Slash-and-Burn Farming"). This is an indigenous skill acquired from generations of experience with localized conditions. Although it took several years before I understood this indigenous knowledge, it was conveyed to me by the actions of indigenous farmers the first time I attempted to establish a research site in the jungle of eastern Peru. During that foray through the jungle, local farmers were hired to cut trails and dig pits to permit examination of the soil. As one pit was being dug, I inquired of the local men whether they considered that a good site to burn and plant crops. After discussing the question, the reply was negative. A few hours later at another pit site, the question was repeated. After a discussion, the indigenous farmer's response was affirmative. I was recording no differences between the soils at each of the sites but had observed that while the men were discussing the merits of each site, they did not give any attention to the soil. Rather they had fixed their attention on the vegetation. Only after obtaining chemical analyses of the soils and several years of experience in the area did I realize the indigenous wisdom displayed by the actions of the indigenous farmers. All the soils were chemically determined to have minimal amounts of available nutrients. By observing the quantity of vegetation available for burning, the indigenous farmers were selecting a favorable site to establish a crop field based on the expected quantity of essential elements that could be obtained from the ash. Areas with big trees that were difficult to cut with the tools available and whose trunks would not burn were less desirable than areas vegetated by a dense growth of smaller trees. The soil was unimportant to their decision.

When populations are low and unlimited areas of land are available, slash-and-burn subsistence farming practices follow a rather predicable sequence of cutting and burning an area of adequate size to plant food crops for the number of individuals in small family or community groups as outlined in Chapter 5. Figure 8–17 shows an aerial view of a typical slash-and-burn family operation in an area where unlimited amounts of land are available.

Specific rotations differ among slash-and-burn cultivators, but usually the first crop planted after burning is the locally most desirable grain crop. Success is reasonably assured by the supply of essential elements released as the cut-and-dried vegetation and organic residues are burned. Depending on the availability of moisture, a legume crop such as beans may immediately be planted following harvest of the first crop. Nitrogen is often the most limiting element due to leaching of nitrate and removal in the first crop, but legumes obtain nitrogen from the air via symbiotic nitrogen-fixing microbes. After the harvest of two crops, most of the essential nutrients made available by burning the biomass and those supplied from the organic pool by rather rapid oxidation of organic matter in the soil have been harvested in the food crops, and weeds, which have been suppressed by the burning prior to the first crop, often become a severe problem. At this time the slash-and-burn cultivators in the tropics often plant a slow-growing crop like cassava (*Manihot esculenta*) or plantains (*Musa paradisiacal*) that can compete with weeds and provide a reliable food, of low nutritional quality, over the next year. That parcel of land is then abandoned for several years as slow-growing native vegetation attains a sufficient size that the cutting and burning can be repeated.

In temperate latitudes where planting is interrupted by cold winter weather, farmers often planted crops for several successive years after clearing and burning until yields declined from a lack of available essential elements and they abandoned the land. In most temperate latitudes where slash-and-burn farming was historically

FIGURE 8–17   Aerial view of an area approximately 100 acres (40.5 ha) cultivated by a family of subsistent farmers using slash-and-burn
technology in the Amazon jungle of Peru, SA. Note that there are
only a few cultivated fields among several former fields in various
stages of jungle regrowth during a *fallow period.*

practiced, many of the areas abandoned by slash-and-burn farmers have now been
reclaimed and successfully farmed with annual applications of fertilizer to restore
depleted nutrients and lime to neutralize acid soil conditions. At present slash-and-
burn practices are mainly confined to lesser well-developed countries in the tropics.

There is no set time for slash-and-burn cycles. A 20-year so-called fallow period
during which native vegetation is allowed to grow appears to be approximately correct
for some areas of slightly fertile soils in the upper Amazon basin. The basic requirement
of successful slash-and-burn farming is the fallow period during which slow-growing vegetation, usually trees, are allowed to grow in soil with such low available nutrient content
that rapidly growing food crops would not produce an economic return to the farmer.
The amount of land needed to sustain a family subsisting from slash-and-burn technology varies. The typical family surviving by slash-and-burn farming in a humid tropical
jungle probably occupies about 100 acres (50 ha). At any given time, food crops are produced on only about 5 acres (2 ha) with the remainder of the land fallowed to natural
growth. With a fallow period of 20 years to allow nutrient uptake by slow-growing trees,
this practice will involve the entire 100 acres (50 ha) if uninterrupted (Figure 8–17).

The fallow period varies from as little as four or five years on soils containing
rather abundant supplies of calcium, phosphorus, and potassium to an indefinitely long
period of time on deep sandy soils where quartz (SiO_2) composes more than 90% of

FIGURE 8–18 Photo of farmers lighting a fire to burn an area of cut-and-dried jungle vegetation in preparation for planting a food crop in Peru, SA.

the solid particles in the soil. The judgment of the individual farmer determines when there is enough biomass that a successful crop will be produced when that biomass is cut and burned. On the more fertile sites, weed growth often severely hampers continued crop growth. A short fallow period that accumulates enough biomass to support a hot fire and kill weed seeds in the surface soil is all that is needed. Careful preparation is involved in conducting a burn prior to planting. The material to be burned must be as dry as possible to assure a hot fire. The lack of a reliable dry season in the most humid areas severely reduces the intensity of the fire and the amount of ash produced. A good farmer carefully distributes the cut vegetation to assure uniform ash distribution and allows it to dry sufficiently to ensure that most of it will burn. On the day of the burn, the entire area that has been cut and allowed to dry is ignited as rapidly as possible to attain maximum intensity of the fire (Figure 8–18). If the distribution of the material to be burned is not carefully done before burning, ash will not be uniformly distributed and uneven crop growth will result (Figure 8–19).

Appreciable amounts of essential elements are lost in a slash-and-burn fire via air currents generated by the fire (Fernandes and de Souza Matos, 1995) (Table 8–3). Although elemental losses attributed to fire are large, three distinct advantages to the planter result. The essential elements remaining in the ash are mainly in soluble plant-available inorganic forms ready for uptake by the crop plant, the pH value in the surface of these normally acid soils is increased, and weed seeds near the surface are killed.

The reduction of weed competition during the growing of the first crop following a fire is a great advantage to the farmer (Mt. Pleasant, McCollum, and Coble, 1992).

FIGURE 8–19   Photo of uneven growth of rice resulting from uneven distribution of ashes in Indonesia. The blackened areas with the better rice plants had an ash deposit during the burn; the lighter colored soil surfaces with poorer rice plants did not.

TABLE 8–3   Range of Nutrient Element Content in Natural Amazon Jungle Vegetation and Amounts Lost During Burning

Element	N	P	K	Ca	Mg
			Lb Ac^{-1*}		
Natural Biomass	89–536	9–36	179–357	134–1005	27–152
			%		
Loss in Burning	88–95	42–51	30–44	33–52	31–40

*Lb Ac^{-1} × 1.12 = kg ha^{-1}.

Source: Fernandes and de Souza Matos, 1995.

One implement that contributes to weed suppression following a fire is the primitive planting stick (Figure 8–20). Slash-and-burn farmers seldom have the tools and power to greatly disturb the soil. Their technique for planting seed is to poke a hole in ground, insert the seed, and cover it with a deft step of a bare foot. Heat from the fire only kills seeds within a few inches of the surface, and this planting technique does not stir up the weed seeds from below that have not been killed by the fire. Many sincere attempts to improve the life of subsistent slash-and-burn families by providing better

FIGURE 8–20 Photo of a subsistence farm family in the jungle of eastern Peru, SA. Note the wooden planting stick used to make a shallow hole in the soil for seed to be planted.

cultivation tools often have created weed problems by stirring up seeds from below the soil surface.

As human population density increases within an area, it becomes more difficult to determine the length of slash-and-burn cycles because the lack of land forces farmers to slash and burn smaller amounts of vegetation from which they derive smaller amounts of essential elements in the ash and therefore have lower yield expectations. Table 8–4 gives examples of the quantity of essential elements contained in the ashes of forests of different ages at two locations in the Amazon basin.

Slash-and-burn operations are conducted with little concern for topography or rocks (Figure 8–21). Because all planting and harvesting is done with hand tools, avoiding a rock is no more difficult than avoiding a large tree trunk that was not completely consumed in the fire. Depending on personal circumstance, some slash-and-burn farmers may rearrange the unburned logs after the removal of their first or second crop and again burn the now drier logs and again plant one crop.

Sincere attempts have been made to relieve farmers of the laborious task of cutting forest vegetation, and individuals and governments have used bulldozers to clear areas for local farmers. But bulldozer clearing creates several problems. The heavy equipment compresses the soil surface, and infiltration rates are severely decreased. The compacted surface soil is difficult for the farmers to plant with hand

TABLE 8–4	**Nutrient Contribution of Ash upon Burning of Primary Forests and Fallow Regrowth at Different Sites in the Amazon Basin**

Location and Soil	*Vegetation*	*Ash*	*N*	*Ca*	*Mg*	*K*	*P*	*Zn*	*Cu*	*Fe*	*Mn*
		Ton Ac^{-1}				*Lb Ac^{-1}**					
Manaus, Brazil (Oxisol)	Primary Forest	4.1	71	73	20	17	5	0.2	0.2	52	2.1
	12-Yr. Secondary	2.1	37	68	23	74	7	0.3	0.1	20	1.2
	5-Yr. Pasture	1.0	16	52	13	36	3	—	—	—	—
Yurimaguas, Peru (Ultisol)	25-Yr. Secondary	5.4	113	155	38	117	15	0.4	0.2	4	9.9
	17-Yr. Secondary	1.8	60	67	14	34	5	0.4	0.3	7	6.5
	11-Yr. Secondary	0.5	9	194	46	72	7	0.6	0.1	2.4	3.0

*Lb Ac^{-1} × 1.12 = kg ha^{-1}.

Source: Fernandes and de Souza Matos, 1995.

FIGURE 8–21	**Photo of corn and upland rice planted in a field among rocks and unburned logs in Brazil, SA.**

tools. Bulldozers remove essentially all organic residues from the cleared area, thus depriving the soil of essential elements, and without fire to kill weed seeds, severe weed problems are usually encountered. Mechanical, that is, bulldozer, clearing is usually more expensive than hand clearing and not well suited for preparing fields for indigenous farmer use. Mechanical clearing has been successful in preparing fields when lime and fertilizer can be incorporated into the soil with powerful cultivating equipment that also loosens the compacted surface soil and chemical weed control is available.

Sustained production of crops has been demonstrated on slightly fertile soils in the humid tropical jungle. In 1972 a study was started in eastern Peru by clearing an area of jungle and establishing a new experiment station. As part of the experiment station a long-term site was established in which part of the site was managed with inputs of lime and fertilizer recommended by soil-testing techniques. An adjacent area on the same type soil was cultivated and planted in the same way but without any lime or fertilizer. Two and sometimes three crops of corn, soybeans, or upland (not flooded) rice were planted each year on both plots. The yields obtained from each crop for the next 20 years are plotted in Figure 8–22. Crop yields in the unfertilized plots (control plots) dropped to zero within three years and remained there until an unintentional application of fertilizer was made in 1987. Plantings in 1988 and 1989 were discontinued because of terrorist activities in the area, and during that time both plots were

FIGURE 8–22 Graph of grain yields in fertilized and unfertilized plots on slightly fertile soils in eastern Peru, SA. Tons $ha^{-1} \times 893 = lb\ Ac^{-1}$; corn and soybeans weigh 60 lb Bu^{-1}; rice weighs 45 lb Bu^{-1}.

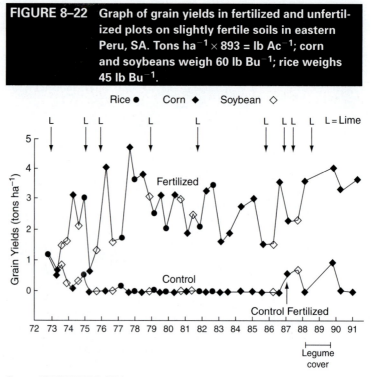

planted to a legume crop. In 1989 one planting of corn produced a small yield on the control plot, but two subsequent crops failed to yield any grain. Note that in Figure 8–19, yields of the first couple of crops on the fertilized and limed plots were not substantially better than on the unfertilized control plots. As researchers were better able to interpret soil test results and adjust quality and quantity of lime and fertilizer to local conditions, reasonable yields were reliably sustained through 39 crops.

CHITEMENTE IN ZAMBIA

Numerous variations of the slash-and-burn scenario just outlined are practiced in tropical forests around the world. In parts of southern Africa, insufficient biomass is present in the native forests to provide the ash necessary for crop production. There planters practice a form of slash and burn called *chitemente*. To accumulate enough biomass, planters cut and gather vegetation, pile it in areas of approximately a half acre (one-fourth ha) to dry and then burn. To minimize labor the gathering and piling of biomass usually results in a circular pattern of bull's-eyes with a cropped area in the center where the biomass was burned surrounded by a larger circle from which the vegetation has been gathered (Figure 8–23).

As population density has increased in many of the traditional slash-and-burn areas of the world, the amount of land available to each farmer has become limited. As

FIGURE 8–23   Aerial photo of bull's-eyes created by *chitemente* practices in Zambia.

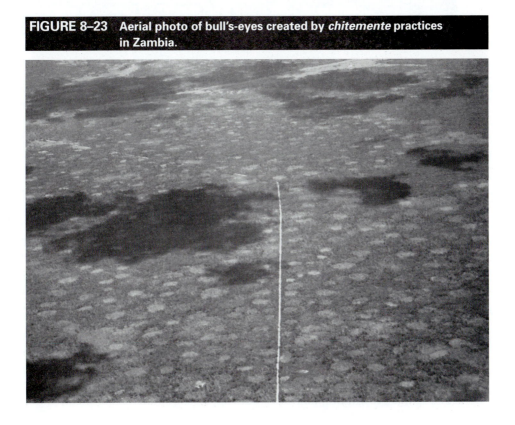

FIGURE 8–24   Photo of slash-and-burn fields in Rwanda. Population density and demand for food are so great that fallow periods of only a few years are permitted on the limited land available.

the need for food increases, the fallow period for jungle regrowth is reduced. As fallow periods become shorter, lesser amounts of available nutrients are obtained upon burning and lower crop yields are obtained (Figure 8–24). The inevitable result of increasing population density and shorter fallow periods in slash-and-burn areas is hungry people who resort to violence, often against adjacent slash-and-burn cultivators of competing ethnic cultures, in attempts to acquire more land.

COMPOSTING

Composting is a common method for concentrating the essential elements in organic residues into material more easily applied as fertilizer to crop fields. The composting process involves the compaction of organic residues and other waste materials into confined compost piles for several weeks or months. During that time microbes partially decompose the organic materials, and much of the carbon is vented to the air as carbon dioxide. Essential elements contained in the materials are concentrated in the resulting composted material and used to improve the fertility of land areas used for food crops.

The composting of dead leaves and grass has been practiced in many parts of the world and is commonly advocated by organic gardeners. The long-term result of gathering vegetation to be composted and used to fertilize food crops can be observed in parts of southern China. In Figure 8–25, the severely eroded hillside vegetated only with a sparse growth of pine trees would appear to be the result of cultivation and a

FIGURE 8–25   Photo of a severely eroded hillside in the southeastern area of the People's Republic of China. Local scientists stated that the hillside was never cultivated, but for many years all plant litter was gathered each year and composted to fertilize paddy rice fields in the valley seen in the foreground.

recent attempt to reforest the eroded land. However, local scientists claimed that the hillside had never been cultivated. Instead local farmers had for perhaps 100 years or more gathered leaves, needles, and other plant residue from the site, composted the material, and spread the compost on adjacent fields in the valley bottom where they grew rice and a few other food crops. By depriving the hillside of the natural recycling of essential elements, the soil had become so depleted that only sparse regrowth of trees resulted, and without vegetative cover to protect the soil considerable erosion had occurred.

The struggle and ingenuity of indigenous people to enhance the chemical fertility of relatively infertile soils has taken many forms in various parts of the world. Composting is one method of concentrating essential elements when human population density exceeds the amount of land necessary to permit enough time to grow sufficient quantities of native vegetation for successful slash-and-burn systems. Although labor intensive, composting is a mechanism to transport essential elements from uneatable vegetation, often on steep landscapes, onto more desirable landscapes and fertilize desirable food crops. The composting process does not create essential elements but only concentrates them in a more available form. The area of land from which the composting material is gathered experiences a loss of essential elements.

MANURE GATHERING

Herding animals is another method of concentrating essential nutrients from native vegetation. Where land area and native vegetation are sufficient for grazing, sheep, goats, cattle, or camels are daily herded into confined areas at night thus permitting the gathering of manure for use on fields cultivated for food crops. Physiological respiration within the animal oxidizes the carbon in the vegetation they consume and concentrates essential elements in the manure, making it a valuable fertilizing material.

Evidence of intense use of manure can be seen in parts of Western Europe where for centuries cattle were grazed during the day and confined to shelters, often in the same shelter occupied by the indigenous humans. Manure mixed with straw and other materials used to bed the cattle was gathered and applied to those areas of surrounding land most suitable for the cultivation of food crops. In parts of Western Europe, this practice apparently began during medieval times and persisted into the early 20th century. These practices created soils with thick black surface layers of chemically fertile soil over the naturally sandy and relatively infertile soil often know as *Plaggen* soils (Figure 8–26). It

FIGURE 8–26 Photo of a Plaggen soil profile with a thick organic rich surface created from decades of manure application in the eastern part of The Netherlands.

Note: Each black-and-white unit of the tape measure is 4 inches (10 cm).

is interesting to note that the land selected for spreading the manure was most often those parts of the landscape where the water table was within about 50 and 100 inches (1.3 and 2.5 meters) below the surface. This assured crops planted on the Plaggen soils of adequate water in the sandy soil material.

MINERAL FERTILE SOILS

The amount of food that can be produced with no attention to external sources of essential elements removed in the human food depends on many variables intrinsic to individual soils. Unlike the naturally less fertile soils, some of the naturally most fertile soils in the world are able to continually sustain low food crop yields for more than 100 years without additions of essential elements.

As discussed in Chapter 6 ("New from the Top Down: Volcanoes and Floodplains"), soils on floodplains and in areas of active volcanoes are distinctly unique in their ability to sustain the harvest of food crops. These areas are of limited extent throughout the world, but where present with appropriate temperature and moisture they play a unique role in human civilizations. Therefore a more complete understanding of how they form and function appears appropriate. Their uniqueness lies in natural processes that add essential element-rich material to their surface.

Floodplains are relatively level areas of land adjacent to almost all rivers that seasonally, or at more infrequent intervals, are covered by water for short periods of time. Floodplains usually maintain a cover of vegetative growth that is not destroyed by the occasional flooding. As floodwaters pass through the vegetation, the velocity is slowed. Suspended particles of soil settle out of the water and cover the previous soil surface. The deposited material is largely free of stones, and the soils formed are well suited for growing food crops. A very consistent and predictable pattern of sedimentation takes place within a floodplain. Near the river channel, where the current is swift during the flooding event, a higher velocity of water is maintained during the flood event and sand-sized particles are deposited. Further from the channel the velocity of floodwater is further reduced by vegetative obstruction and the much smaller clay-sized particles are deposited. Thus a sand-rich zone, called a *levee,* is created near the river channel and more clay-rich sediments are deposited further from the river channel. The sandy levee may build by additions of material during each flood and eventually be several feet above the land further from the river. During an intense flood the water may destroy a part of the natural levee and erode a new channel, leaving the old channel to fill with sediment in succeeding floods (Figure 8–27). Successive flood events during hundreds and thousands of years form a complex pattern of sandy and clayey soils within floodplains.

A common feature of all floodplains is that the soil surfaces receive new soil material with each flood event. Much of the soil material deposited during the floods has been derived from the erosion of topsoil throughout the watershed and contains considerable amounts of organic residue rich in nitrogen and other plant-essential elements.

Floodplains are adjacent to rivers that can provide salt-free water for irrigation in areas where rainfall is too low for crop production. The importance of essential element renewal and irrigation water on floodplains is evident in historical development and intensification of civilizations on floodplains and deltas throughout the world.

FIGURE 8–27 Aerial photo of a meandering river in the upper Amazon basin in Peru, SA. The river has changed course and left a horseshoe lake in its former channel.

Proximity to rivers also provided floodplain areas easy access to transportation even prior to the invention of the wheel. If we think back into the history of civilization, we find the first evidences of city development are on the floodplains of the Nile, Euphrates, and Tigris Rivers in the Near East. Dense populations today occupy and endure frequent floods on the floodplains of the Ganges, Mekong, and Yellow Rivers of Asia. As we see the suffering these floods cause, we may ask, "Why do people live there?" Certainly, part of the answer is the reliable natural renewal of chemical fertility that permits sustained crop production. Although primitive people did not understand the chemistry involved, they found that each year they were assured of being able to grow food crops. People who survived the floods were assured of nutritious food, grown on fertile soil. The physical dangers of severe floods were a more acceptable risk than the uncertainty of food on less chemically fertile land.

The detrimental effects of flooding have caused humans to attempt protection of many floodplains with dams and levees. Dams and levees restrict the natural restoration of essential elements via flooding, and modern methods of fertilization are now utilized in many, but not all, floodplains of the world (Figure 8–28).

There are many relatively uninhabited remote areas where flooding is not controlled. One such area is in western Venezuela on the plains known as Llanos where large areas of fertile land are seasonally flooded by runoff water from the Andean

FIGURE 8–28 Aerial photo of intense paddy cultivation in the Bangkok Plains of Thailand.

mountains to the west (Figure 8–29). The many rivers that flow from the mountains have thus far defied attempts to control the flooding and the building of reliable roads. Most of the land is use to graze cattle.

Perhaps a more unique natural mechanism of replacing essential elements harvested from cropland is volcanic activity. Molten material ejected from volcanoes quickly cools and solidifies in the air forming fine-textured ash particles. These ash particles rapidly weather when subjected to conditions at the earth's surface. If the volcanic material is rich in plant-essential elements, the ashy dust that settles around the volcanic vent is a source of essential elements. All volcanic materials are not rich in essential elements, and certainly the climatic conditions of temperature and rainfall necessary to support crop growth must be present to augment the chemical fertility of the volcanic ash if the resulting soil is to be considered fertile. Dense populations in Japan, Java, parts of Italy, the Rift valley in Africa, and the volcanic mountains of Central and South America are examples of human attraction to volcanic activity for support of intense human populations.

FIGURE 8–29   Aerial photo of flooded land (Llanos) in southwestern Venezuela, SA.

Volcanic ash is often present on steep mountain slopes. Seeking the long-proven fertility of soils formed in volcanic materials, many such slopes are intensively cultivated. Most soils formed from volcanic ash are friable and have a high infiltration capacity such that runoff on steep slopes is less than on most other soils. But, as seen in Figure 8–30, soils formed from volcanic ash on steep fields are not immune from erosion, in this case small landslides in the farmer's field.

Volcanic areas have supported indigenous food crop production for many years. The first essential element to become deficient is nitrogen. The growing of legumes to fix nitrogen from the air in annual rotations with food crops was known in Roman time and rediscovered, as it were, on the fertile soils in England between 1650 and 1850 (Loomis, 1984). Although the improvement of crop yields is hampered by the capacity of the oxides in most soils formed in volcanic ash to fix additions of soluble phosphate fertilizers, there is a rapid enough release of essential elements to provide reliable production of food crops, albeit the yields often are rather low. The long-term reliability of food production is witnessed by the vertical orientation of fields resulting as land ownership was divided among heirs from generation to generation on volcanic ash land in Guatemala (Figure 8–31). Although subjected to tragic human disasters during sporadic and unpredictable volcanic eruptions, humans flourish near many volcanoes. Although their lungs may be subjected to injury from volcanic dust, their diets are rich in essential elements from food grown on these soils.

Naturally fertile soils with adequate periods of warm temperatures and rainfall on gently sloping land in Europe and North America were highly prized for their ability to produce food crops and are now almost entirely devoted to food crop

FIGURE 8–30 Photo of intensely cultivated fields on steep slopes in Ecuador, SA. Small landslides are evident in some fields.

production as seen in Figure 8–32. The natural fertile soils in these areas have resulted from the deposition of crushed limestone with admixtures of other rocks from continental glaciers. The most favorable soils have developed in thick layers of silt-sized particles deposited as loess. The chemically fertile silt and silt loam textured soils that formed in the loess also have a high available water-holding capacity that assures moisture to grain crops during the warm summer months. As discussed further in Chapter 9, these soils have been capable of sustained crop production for many years without additions of fertilizing materials and are often considered the best soils in the world.

PERSPECTIVE OF ACTIVITY ON THE LAND

Land and soil as we observe them are not permanent in a geologic time scale. Changes in land and soil result from both imperceptibly slow processes and such infrequent catastrophic events that they often escape human recognition. Natural ecosystems of flora and fauna seek a niche on the land where they have an advantage over other types of flora and fauna. Temperature, water, and availability of essential elements determine where that niche is on the land. Humans have also sought their niche on the land and multiplied where their ability to produce food reliably made habitation possible. Cold temperatures constrain human habitation except insofar as food can be transported. For many generations humans have been able to alter water constraints on some arid lands and wetlands with irrigation and drainage technology. Where

FIGURE 8–31   Photo of vertical land ownership pattern of fields in the volcanic mountains of Guatemala, CA.

water and temperature conditions are, or can be modified to be, compatible with human food production, the search for available essential elements in the soil has dictated the distribution of human habitation. Human ability to understand and then compensate for inadequate chemical fertility in the soil is rather recent in human history. Utilization of modern technologies to enhance chemical fertility has profoundly affected food crop production in many areas, but many and perhaps the majority of the people on earth are not able to avail themselves of these technologies. Chapter 9 explores the advantages and constraints to utilization of technology in growing food crops.

FIGURE 8–32   Aerial photo of the intensely cultivated fertile soils in the Midwest of the United States.

LITERATURE CITED

Fernandes, E .C. M., and J. C. de Souza Matos. 1995. Agroforestry Strategies for Alleviating Soil Chemical Constraints to Food and Fiber Production in the Brazilian Amazon. In P. R. Seidl, O. R. Gottlieb, and M. A. C. Kaplan (Eds.), *Chemistry of the Amazon: Biodiversity, Natural Products, and Environmental Issues* (pp. 34–50). ACS Symposium Series 588. Washington, DC: American Chemical Society.

Haverstock, N. A. 2004. Brazil. In *Encyclopedia Year Book 2004* (pp. 101–102). Grolier., USA.

Judson, S., and D. F. Ritter. 1964. Rates of Regional Denudation in the United States. *Journal of Geophysical Research* 69:3395–3401.

Lehmann, J., D. C. Kern, B. Glaser, and W. I. Woods 2003. (Eds.), *Amazonian Dark Earths* (p. 505). Dordrecht: Kluwer Academic Publishers.

Loomis, R. S. 1984. Traditional Agriculture in America. *Annual Review of Ecology and Systematics* 15:449–478.

Lopes, A. S. 1996. Soils Under Cerrado: A Success Story in Soil Management. *Better Crops International* 10(2):9–15.

Mt. Pleasant, J., R. E. McCollum, and H. D. Coble. 1992. Weed Management in a Low-Input Cropping System in the Peruvian Amazon Region. *Tropical Agriculture (Trinidad)* 69:250–259.

TROPSOILS. 1991. *Annual Report.* North Carolina State University Soil Science Department, Raleigh.

Wright, A. C. S., and J. Bennema. 1965. *The Soil Resources of Latin America.* World Soil Resources Office, Land and Water Development Division, Food and Agriculture Organization, and United Nations Educational, Scientific and Cultural Organization, Rome.

NOTES

1. See Chapter 1 of E. J. Russell, *Soil Conditions and Plant Growth* (8th ed.) (London: Longmans, Green and Co., 1950), for a review of the earliest writings.
2. Land prices in the *cerrado* at that time were approximately US $50 per acre. Total cost to buy, clear, lime, and fertilize 1 acre was approximately US $300, and fertile land in the Midwest of the United States cost approximately $3,000 per acre.

CHAPTER REVIEW PROBLEMS

1. What does geologic material contribute to the fertility of the land?
2. Under what conditions is more material lost from the soil by dissolution than by erosion?
3. Why does vegetative growth rate usually increase after a fire destroys existing vegetation?
4. With so much of the essential elements in the biomass lost during burning, why do slash-and-burn farmers burn?

CHAPTER

9

What It Takes to Do Our Job

Land is space for people to live and grow most of the food they need for life. Without food that decomposes within our bodies to give us energy, no person can live very long. There are many potential food materials, but only a few have found favor as food sources for humans. We are all familiar with some of fruits and vegetables that frequent our dinner plates. The grains like corn, soybeans, wheat, and rice are less identifiable because we usually only see them in the form of bread, pasta, cake, and cookies. We are all familiar with meat as a food but perhaps do not consider that the animals that appear on our dinner plates as beefsteak, pork chops, and chicken legs consumed grain, grass, and other plants. Fish and other seafood grow at the top of an aquatic food chain. That food chain begins with algae and other organisms that require nitrogen, phosphorous, calcium, magnesium, potassium, and other essential elements dissolved in river, lake, and ocean waters but derived in large part from erosion and leaching of soil on surrounding watersheds. Although it may require taking a few steps back into the life cycle of some foods, all human foods have some common requirements that can be traced to processes on land and in soil.

HARVESTING SUNSHINE

Sunshine, and the warm temperatures that result in its presence, are direct or indirect requirements for all life on earth. The amount of energy that can be obtained from sunshine depends on the duration and intensity of the sunshine and size of the area available to absorb radiation from the sun.

The duration of sunshine has two aspects of concern in food crop production. One is day length and the other is number of consecutive days a particular crop requires to mature. The proportion of each day during which sunlight directly radiates the land surface varies with north-south location on the earth. At the equator each 24-hour day contains 12 hours of sunlight and 12 hours of darkness, every day of the year. As we move north or south of the equator we have a summer season in which daylight exceeds 12 hours, and conversely in the winter season daylight is less than 12 hours each day. The magnitude of these seasonal differences becomes greater the further we move from the equator. From the Arctic and Antarctic circles to the respective poles, brief summers have 24 hours of sunlight each day and in winter no sunlight hours.

We all have experienced that the intensity of sunshine is greater during clear days than during cloudy days and greatly reduced in shaded areas. Perhaps we have seen the effect of shade on the reduction of crop growth or experienced poor growth of grass in our lawns under the shade of a tree. Almost all plants grow more slowly in shade than in full sunlight, although some species can grow well with some shading. Subsistence farmers using intensive hand labor to cultivate and harvest the food crops often utilize the practice of planting more than one crop species at the same time in the same field. This practice is often referred to as *intercropping*. Some yield reduction of the smaller plant species results because of shading, but intercropping of food crops provides the subsistence farmer with a variety of food and reduces the risk of complete crop failure from disease or insect damage that may affect specific crops. Although a high yield of food is often attained with intercropping, most mechanical methods of harvest are impossible when various crops of different size and shape and different times of ripening are mixed on the same location. The planting of a single species, *monocultures,* of food crops has become an almost universal practice in mechanized commercial farming.

The vast majority of humans live in temperate and tropical latitudes. Generations of humans have evolved a myriad of technologies to produce human food directly from plants and from animals that also consume plants. In recent generations science has genetically altered many of our most important food crops to perform best with long summer days in the temperate latitudes. Hybridization and other plant-breeding techniques have been able to reduce the growth cycle of many crop plants so that mature grain can be produced within the seasonal temperature limitations of temperate latitudes. Several physiological processes within plants are controlled by day length, and some cultivars adapted to temperate latitudes grow poorly or fail to produce grain when grown in the shorter day lengths of the tropics. Many crop species have been genetically selected to ripen uniformly, thus accommodating mechanical harvest.

These and other genetically controlled aspects of plant growth have helped increase food crop production. However, all plants require radiant energy from the sun. There is a limit to how densely crop plants can be spaced because of mutual shading. Therefore, to assure crop plants adequate access to radiation from the sun, the amount of land devoted to crop production is ultimately determined by the amount of food required and the efficiency (i.e., crop yield per unit of land) with which that food can be grown. To increase the total production of human food, there are two obvious pathways: either increase the area of land planted to food crops simply to increase the total harvest of sunshine or increase the yield of food crops per unit of land.

SIZE OF THE "SOLAR PANEL" CROPLAND

How much land must be planted to food crops to meet the food requirements of increasing human populations? History of food production in the United States demonstrates that the simple assumption that the presence of more people requires more land to be cultivated for food crop production is not the only answer. In Figure 9–1 we see that population in the United States increased each 10-year period from 1910 to 1990. We also see the total number of acres harvested for food crops in each of those years and the number of acres harvested to feed only the people within the United States. The difference between total harvested acres and acres harvested for U.S. consumption represents the number of acres used to grow

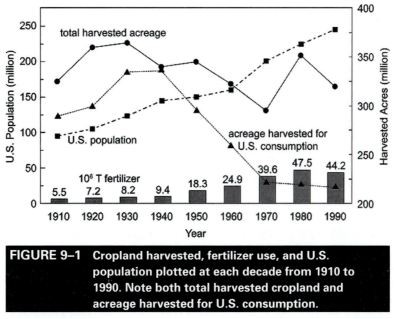

FIGURE 9–1   Cropland harvested, fertilizer use, and U.S. population plotted at each decade from 1910 to 1990. Note both total harvested cropland and acreage harvested for U.S. consumption.

Source: Buol, 1995.

food crops for export to other countries. Note that from 1910 to 1940 the number of acres harvested for U.S. consumption increased at about the same rate as the population increased. During that period there was a substantial export of food, primarily to Europe, which almost disappeared in 1940 because of the outbreak of World War II and the disruption of shipping in the Atlantic Ocean. Between 1940 and 1970, there was a drastic reduction in the number of acres harvested for U.S. consumption and a dramatic increase in exports. Population in the United States increased steadily throughout that period, but the amount of cropland producing food for U.S. consumption dramatically decreased. Clearly, the yield of human food per unit area of land greatly increased during this period. Meeting the human food demand from less cropland has been a dramatic departure from previous experiences of civilization. It is not a condition experienced in all parts of the world today.

No single factor is responsible for the dramatic reduction in the amount of land required to produce food for each U.S. citizen from 1940 to 1970. One of the major factors is illustrated by the bar graphs at the bottom of Figure 9–1. Farmers in the United States greatly increased their use of fertilizers after World War II, thereby decreasing the chemical limitations to plant growth and increasing the grain yield per acre. Note also that during this time great improvements were made in plant genetics that increased the physiological capacity of the crop plants to produce grain, *if* there was a sufficient supply of essential elements in the soil. An often unheralded contribution to the increased food production per unit of land was more efficient transportation and refrigerated storage of perishable foods between the farm and the dinner table. With less spoilage, a greater proportion of the food produced on the farms reached the dinner plate. During this period of time, improved control of plant and animal diseases and weeds became available through the use of new pesticides. Also, immediately after the end of World II, larger and faster tractors and tillage equipment became available to farmers in the United States. Each farmer was able to plant, cultivate, and harvest more acres per day. This ability to do necessary planting, cultivating, and harvest operations in a timely fashion, that is, when weather conditions are favorable, is a component of food production that is often overlooked. There are only a limited number of days during each year when the soil moisture content is suitable for tillage and/or harvesting. The art of farming, simply stated as doing things "on time" with "Ole' Man Weather" in charge of the clock, is a major factor in attaining high crop yields per unit of land. Executing farming operations on time is a characteristic of all good farmers. The more powerful equipment enabled the most skilled farmers to apply their management skills to larger areas of land and resulted in increased size of individual farms and a lesser number of farmers in the United States.

Although we have found soil useful to support buildings, roads, and airports, its role of holding plants upright so they can capture the radiant energy of sunlight and ingest the carbon dioxide contained in the air and water from the soil are primary functions of soil. Efficient high-yielding cropland requires that native vegetation be removed to eliminate competition for sunlight and available soil water. Animals that grazed the native vegetation must also be removed to prevent their consumption of the food crops (Figure 9–2). The buffalo herds that once grazed the native grasslands in the United States and Europe are not compatible with fields of corn, wheat, and other food grains.

Humans have been very selective in what kind of land they cultivate for food crops. The lands they have selected for cultivation of food crops differ in many respects, but all

FIGURE 9–2   **Photo of bison (buffalo) in South Dakota, USA. Buffalo herds are not compatible with corn or wheat fields.**

possess properties that people can manage for successful crop production. The ability of people to manage land for food production differs not only with the skill of the individual farmer but perhaps more with the skill of the society that surrounds the farmer. The skill of a society includes the economic, social, and political aspects of the surrounding society to make available the tools that farmers use to grow food crops efficiently. These tools are not limited to the visible hoes or tractors but include the availability of marketing systems in which the farmer can reliably market food products and obtain money to purchase materials used to cultivate the land and harvest the food crops. Other tools include such items as genetically viable seed of the adapted food species and fertilizer, so the crops are not lacking for the chemical elements needed for growth and grain production. Where soils are frequently subjected to insufficient rainfall during the growing season, reliable supplies of irrigation water are needed. If the soils are frequently saturated with water, thus depriving the plant roots of oxygen needed for respiration, drainage ditches to remove excess water are needed. Pesticides to combat crop damage by insects and diseases need to be available at critical, often unpredictable times during the growing season. To grow most food crops successfully, weeds that compete for sunlight and soil-stored water must be controlled. Individual farming skill is not able to provide most of these tools. Irrigation systems require access to water and power to move that water. Drainage systems needed to remove salts from irrigated land require access to areas where the salty water can be discharged. Farmers do not control efficient marketing and transportation systems or the production of fertilizer and tillage equipment. Technologies to control insects and diseases are not developed or maintained by individual farmers, although correct application at the right place and at the right time is a critical farming skill.

All of society carries the responsibility for the success or failure of efficient food crop production. Political stability is fundamental to all sustainable land use for food crop production. Political stability must assure that markets and land tenure are secure so talented individuals can acquire the art of farming skills. Broken bridges, unstable financial systems, lack of quality control in fertilizers and pesticides, and foreign trade policies that interrupt markets are as disastrous to successful agriculture as soil acidity, salinity, floods, or hailstorms.

AGRICULTURE IS BIGGER THAN FARMING

In the preceding discussion, a distinction between farming and agriculture has been implied. *Farming* is used to describe the direct contact between persons and the land. *Agriculture* is much more inclusive. Agriculture includes farmers and all the people involved in providing materials needed by farmers and the facilities needed to get the food from the land to the dinner table. To understand this concept of agriculture, you could pose the following questions: Who do you think makes the tires for the tractors? Who do you think loans money to buy seed and fertilizer? Who do you think develops the hybrid seeds? Who do you think drives the trucks, trains, and boats that transport food? Who do you think wraps fruits with plastic wrap in the supermarket? Are all of these people agricultural workers? At present approximately 2% of the population in the United States are farmers, and almost 20% of the population earns salaries from agricultural and food-related enterprises.

Figure 9–3 is a pictorial representation of the requirements for agricultural production. Total agricultural production of food is represented as a lake level maintained

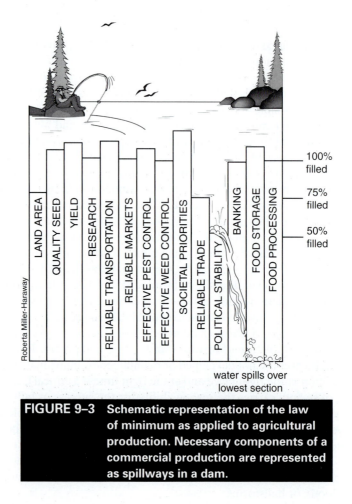

Roberta Miller-Haraway

LAND AREA
QUALITY SEED
YIELD
RESEARCH
RELIABLE TRANSPORTATION
RELIABLE MARKETS
EFFECTIVE PEST CONTROL
EFFECTIVE WEED CONTROL
SOCIETAL PRIORITIES
RELIABLE TRADE
POLITICAL STABILITY
BANKING
FOOD STORAGE
FOOD PROCESSING

100% filled

75% filled

50% filled

water spills over
lowest section

FIGURE 9–3 Schematic representation of the law
of minimum as applied to agricultural
production. Necessary components of a
commercial production are represented
as spillways in a dam.

by a dam. Each spillway of that dam is labeled to represent components required for
agricultural production. The lowest level of any one individual spillway controls the
lake level, that is, total agricultural production. Like the law of the minimum depicted
in Figure 4–1 that represents elemental requirements for the growth of plants,
Figure 9–3 represents the law of the minimum for agricultural production. The figure
contains one spillway labeled "Reliable Trade." That concept can be applied at many
size scales. Depending on the scale to which the figure is applied, it is important to recognize that many countries and also areas within countries depend on importing food
and exporting nonfood products. Farmers require reliable and timely delivery of supplies as well as reliable markets. Implicit in reliable trade is a stable monetary system of
banks and trade agreements.

Within any area only a limited amount of land is suitable for farming. Often only
specific kinds of food crops are adapted to the area. Some countries do not have enough
suitable land area to feed their people and must rely on importation of food. Societal priorities determine whether land has more value for food production, aesthetics, shopping
centers, or urban housing. Total agricultural production depends on not only the amount

of land but also the yield of food crops per unit of land. Although the skill of individual farmers is critical to attaining high yields, many components contributing to high yields are controlled by the societies that surround the farmers. Among the factors of agricultural production beyond the control of individual farmers is the availability of viable seed, fertilizer, and pesticides for controlling diseases and pests that can destroy part or all of a food crop. There must be reliable transportation systems that both convey supplies to the farmers and food crops from the farmers to the consumers. Transportation systems require trucks, trains, and sometimes boats, and maintenance of roads, rails, and waterways for reliable operations. Because most food crops are seasonally produced and demand for food is nearly constant, reliable facilities to store and process food are required. Another imperative is that farmers must be reasonably assured of a market before they will make investments of money for seed and fertilizer in an attempt to grow more food than their immediate family can consume. A stable banking and monetary system is necessary for farmers to invest in the production of food for sale. When there is not reasonable assurance that a market will be available some months and even years in the case of animal and tree crop production in the future, farmers resort to growing only what is needed for their families (i.e., subsistence farming), rather than commercial production. To attain the most efficient production, agricultural research is required to continually fine-tune the selection of cultivars and management techniques to local soil conditions and transient markets. Essentially all of these requirements depend on political stability and the priorities of the entire society present in the area of concern.

Harvesting the energy of the sun to convert the life-essential elements from air and soil into human food at the dinner table is the most fundamental endeavor of humanity. Human population determines the total amount of food required, but how efficiently that food is grown and distributed is determined by the agricultural skill of the entire society. The combination of population requirements for food and efficiency of food production per unit of land determine the amount of land devoted to food crop production.

NO FREE LUNCH: HARVEST EQUALS INPUTS

What do humans want from the land? Perhaps the answer may be a house site, a forest for hiking or hunting, or a place to drive on the open highway, but reduced to the fundamentals of life an affordable source of food would have to top the list of human necessities. Let's dig a little deeper into this question and examine what we require in human food. Good taste, variety, and lack of health-endangering substances are important, but primarily we need nourishment. Although we need that nourishment in the form of protein, vitamins, carbohydrate, fat, and so on, each of these food components is composed of chemical elements. As discussed in Chapter 4, these essential elements naturally originate in air, water, and minerals present in soils. They find their way into human food through growing plants.

The elemental composition of the human body, Table 4–1, compared to the composition of soil, presents part of the picture of how elements have to be concentrated in the food chain to satisfy human requirements. The degree to which essential elements must be concentrated from soil to human bodies gives us a clue to which elements are potentially most limiting. Note that of all the elements whose natural source is in the soil, phosphorus is approximately five times as concentrated in plant tissue as it is in soil minerals and ten times as concentrated in the human body as it is in soil minerals.

To obtain a more pragmatic overview of phosphorus in the food chain, we need to consider the daily requirement for each human. Recommended dietary requirements for phosphorus change with human age. Young children require about 800 mg of phosphorus per day, young adults about 1,200 mg, and older adults about 800 mg per day. If we settle on 1,000 mg of phosphorus per day as a basic requirement and multiply that value by 365 days per year, we arrive at 365,000 mg per year per person (0.803 Lb P Yr^{-1} $Person^{-1}$). Taking a worldwide view and estimating the world population at 5.5 billion people, approximately 4.4 billion pounds of phosphorus per year (2 billion kg P Yr^{-1}) needs to reach the dinner tables of the world. The extraction of this amount of phosphorus from the 3.7 billion acres (1.5 billion ha) of cultivated land in the world is approximately 1.2 Lb P Ac^{-1} Yr^{-1} (1.34 kg P ha^{-1} Yr^{-1}). This calculation makes the unreasonable assumptions that people only consume their minimum daily requirement and no loss of phosphorus occurs as the food is processed and transported to the dinner table.

In Table 5–1 natural soils were selected to illustrate the naturally occurring range of both plant-available phosphorus and total phosphorus contents. We find that soil test, often called "available" phosphorus in a 7-inch (18-cm) thick topsoil layer ranges from 3 to 25 Lb P Ac^{-1} (3.3–28 kg P ha^{-1}). Lesser amounts are almost always present in subsoil layers. As you will recall from the discussion in Chapter 5, some of the more unavailable mineral forms of phosphorus will slowly become available as the minerals decompose. Therefore, if we examine *total* phosphorus contents in soils (Table 5–1) we find values to range from less than 200 to as much as 2,000 pounds per acre in a 7-inch-thick layer of topsoil (224–2,240 kg^{-1} ha^{-1} 18 cm^{-1}). If *all* the phosphorus in a soil could be removed by crops, it is clear that in any soil all of the phosphorus would be removed within a finite number of years just to meet the human requirements for phosphorous. For example, if the minimum amount of phosphorus, 1.2 Lb P Ac^{-1} Yr^{-1} (1.34 kg P ha^{-1} Yr^{-1}) were extracted from a soil with a total content of only 200 pounds of P per acre (224 kg ha^{-1}), the total amount of phosphorus would be exhausted in 167 years. Of course, the assumption that plants could remove all the phosphorous in a soil is unrealistic. As the total amount of phosphorus in the soil decreases, the weathering rates that convert mineral pool phosphorus to available pool phosphorus are so slow that amounts of available phosphorus would decrease to levels where most plants would refuse to grow well before all of the total phosphorus was extracted. Similar calculations could be developed for each of the other essential elements that have to be obtained from soil minerals, but in most soils phosphorus is the most compelling element.

Another way to visualize the impact each person has on the flow of elements is to consider the minimum daily phosphorus requirement over a person's life. Assuming a person has consumed only the minimum daily requirement of phosphorus; by age 80 approximately 64 pounds (28 kg) of phosphorus would have been consumed. The human body contains 1% phosphorus (Table 4–1) and thus approximately 2 pounds (1 kg) of phosphorus would be in a 200-pound (89-kg) person. Thus we can calculate that during the 80-year lifetime a minimum of 62 pounds (28 kg) of phosphorus have been voided in the body wastes of one individual. The effect human populations have on spatially concentrating phosphorus, and other essential elements, from land where their food is grown to waste disposal areas in cities where people are concentrated is obvious.

The daily human requirement for calcium is about the same as that for phosphorus. The daily human requirement for potassium is approximately three times as great as that for phosphorus or calcium. By referring to Table 5–1 you will note that although the quantities of potassium in most soils are greater than phosphorus, exhaustion of potassium would also take place in soils where food crops are harvested without an external source of potassium being supplied. To provide for sustained human food production the choices are either fertilize cultivated lands with external sources of phosphorus and the other essential elements naturally obtained from soil minerals or abandon cropland as the content of essential elements decreases from continued harvest and exploit new land for cultivation.

In most societies the food chain involves human consumption of meat. Conversion of essential elements from the vegetative crop used as animal food to meat introduces further spatial concentration of elements removed from the soil. Approximately 80% to 90% or more of major nutrient elements in the grasses and grains consumed by animals grown for meat end up in animal manure. When animals are allowed to graze vegetation freely, much of the manure is returned to the site where the plants grew. This is seldom possible, especially in temperate areas of the world where animals are not able to graze during the cold months of the year. Some spatial concentration of animal manure is unavoidable in domestic meat and milk production. Greater efficiencies in labor and live weight gain are obtained when meat animals are confined in feedlots and fed controlled diets. In many agricultural systems the concentration of cattle, hogs, chickens, and turkeys has added greatly to the spatial concentration of nutrient elements. There are a number of management alternatives to obtain life-essential nutrients from the soils in the croplands of the world. Perhaps the least desirable option is to continue the harvest of crops until the nutrient-depleted surface soil can no longer support a complete vegetative cover. At that point the unprotected soil surface erodes more rapidly, thereby exposing deeper layers to exploitation by plant roots. To some extent this process has taken place throughout the development of civilizations and led to the abandonment of unproductive and eroded land. To escape this scenario, many people have endured the dangers of living in floodplains and in the shadow of volcanoes where natural events recharge the surface of the soil with a supply of essential elements (Chapter 6, "New from the Top Down: Volcanoes and Floodplains"). Several human cultures have physically carried fertile soil material from floodplains to higher ground, free from flooding, and thereby supplied the elements needed in their food-growing fields. If the growing and export of food crops from one location are to be sustained, the life-essential elements in the soil must be replenished.

REPLENISHING THE ELEMENTS HARVESTED IN FOOD

Perhaps a universal answer to replenishing the elements we harvest in food crops can be found in the complete recycling of all human and animal wastes to the site where the food is grown. This is attempted in many societies with mixed results (Figure 9–4). One clearly negative feature of this practice is the potential for the spread of disease in untreated sewage and manures. In more metropolitan societies, human waste becomes mixed with undesirable and potentially toxic chemicals and heavy metals that may enter the crop plants and impair the health of persons eating products grown where waste materials are used to enrich the soil. Perhaps we can do a better job of sorting our urban waste products to reduce this potential.

FIGURE 9–4 Photo of a woman dipping water from a night soil and compost solution pit to be applied to a vegetable garden in the People's Republic of China.

There is one serious constraint to recycling organic waste that appears unavoidable and endemic to all societies. The content of essential elements in most waste materials is such a low percentage of the total weight that very high transportation costs are encountered relative to the fertilizer benefit from the nutrient elements contained in the waste. The more distant the site of production is from the site of food consumption and waste generation, the greater the cost. Table 9–1 gives examples of essential elemental contents in various waste products available in India.

TABLE 9–1 Average Nutrient Contents in Selected Organic Manures and Plant Residues (% by Weight)			
Material	*N %*	*P %*	*K %*
Farmyard manure	0.5–1.5	0.18–0.35	0.4–1.6
Urban compost	1.0–2.0	0.44	1.2
Rural compost	0.4–0.8	0.13–0.26	0.58–0.8
Green manure	0.5–0.7	0.04–0.08	0.5–0.66
Night soil	1.2–1.3	0.35–0.44	0.3–0.4
Human urine	1.0–1.2	0.04–0.09	0.17–0.24
Cattle dung + urine	0.60	0.07	0.37
Sheep dung + urine	0.95	0.15	0.8
Poultry manure	1.0–1.8	0.62–0.79	0.8
Horse dung + urine	0.70	0.11	0.46

TABLE 9–2A Comparison of Manures, Composts, and Chemical Fertilizer Composition

Fertilizing Material	N, P, K Content (Wt %)			Lb* Material/Lb N, P, K		
	N % Wt.	P % Wt.	K % Wt.	Lb /Lb N	Lb/Lb P	Lb/Lb K
Farmyard Manure	1.00	0.25	1.00	100	400	100
Urban compost	1.50	0.44	1.20	67	227	83
Rural compost	0.60	0.19	0.65	167	526	154
Green manure	0.60	0.06	0.61	167	1667	164
Night soil	1.20	0.39	0.35	83	256	286
Poultry manure	1.40	0.71	0.80	71	141	125
17–17–17 Fertilizer	17.00	7.48	14.10	6	13	7
20–10–5 Fertilizer	20.00	4.40	4.15	5	23	24

*2.2 lb = 1 kg.

The amounts of selected waste products listed in Table 9–1 needed to provide one pound of nitrogen, phosphorus, and potassium are given in Table 9–2A. The weight of two commonly available fixed analysis chemical fertilizers, 17–17–17[1] and 20–10–5, are included for comparison with the various waste products. Table 9–2B presents the weight of each material needed to supply the nitrogen, phosphorus, and potassium harvested in 80 bushels of rice grain and 2.5 tons (T) of rice straw (see Table 5–4) that grows to support that amount of grain. The calculations are for 80 bushels (Bu) of rice and 2.5 T of straw regardless of the amount land used to grow that crop. A bushel of rice grain weights 45 pounds (20 kg). An 80 Bu Ac^{-1} or 3,600 Lb Ac^{-1} (4,032 kg ha^{-1}) yield of rice is a reasonable yield expectation in much of the developing world but low compared to the 2003 average yield of 6,578 Lb Ac^{-1} (7,367 kg ha^{-1}) in the United States (*Rice Journal*, 2004). Table 9–2C lists the amount of material needed to fertilize both the grain and straw that would be the case where the straw from the previous

TABLE 9–2B Amount of Material Needed for Rice Grain and Straw

Fertilizing Material	Lb Material Needed for 80 Bu[†] of Rice Grain[‡]			Lb Material Needed for 2.5 T Rice Straw		
	N	P	K	N	P	K
Farmyard manure	**5,000**	3,600	800	3,000	2000	**5,800**
Urban compost	**3,333**	2,045	667	2,000	1,136	**4,833**
Rural compost	**8,333**	4,737	1,231	5,000	2,632	**8,923**
Green manure	8,333	**15,000**	1,311	5,000	8,333	**9,508**
Night soil	**4,167**	2,308	2,286	2,500	1,282	**16,571**
Poultry manure	**3,571**	1,268	1,000	2,143	704	**7,250**
17–17–17 Fertilizer	**294**	120	57	176	67	**411**
20–10–5 Fertilizer	**250**	205	193	150	114	**1,398**

[†]One bushel of rice equals 45 lb (20 kg).

[‡]2.47 Ac = 1 ha.

TABLE 9–2C Amount of Material Needed for Total Rice Crop

Fertilizing Material	Lb Material Needed for Total Crop		
	N	P	K
Farmyard manure	**8,000**	5,600	6,600
Urban compost	5,333	3,182	**5,500**
Rural compost	**13,333**	7,368	10,154
Green manure	13,333	**23,333**	10,820
Night soil	6,667	3,590	**18,857**
Poultry manure	5,714	1,972	**8,250**
17–17–17 Fertilizer	**471**	187	468
20–10–5 Fertilizer	400	318	**1,590**
20–8–20 Fertilizer Blend	**400**	**400**	386

crop had been removed. A 20–8–20 custom-blended fertilizer available from modern fertilizer supply facilities is created for comparison.

Obeying the law of the minimum the weight of each material needed to provide the amount of each element taken up by the rice crop was then calculated. In Tables 9–2B the amounts needed by the grain and straw are calculated separately. In Table 9–2C the amount of each fertilizing material to adequately supply *all* of the elements in both the straw and grain of an 80-Bu rice crop is calculated. The element for which the greatest amount of each fertilizing material is required is presented in **bold** type. The lesser amounts of fertilizing material needed to supply each of the other elements adequately are listed in regular type. The difference between the amount of fertilizing material identified in **bold** type and the amount needed for each of the other elements identifies how much of that element would be added in excess of the crop requirement to assure an adequate supply of *all* essential elements. For example, in Table 9–2B 15,000 pounds (6,818 kg) of green manure is required to provide the phosphorus necessary, whereas only 8,333 pounds (3,788 kg) of green manure would be required to provide enough nitrogen in the grain. Thus adding the additional 6,667 pounds (3,024 kg) of green manure containing 0.6% nitrogen to provide the necessary phosphorus would supply 40 pounds (18 kg) more nitrogen than needed by the crop. The excess nitrogen is a potential source of nitrate contamination in the surrounding environment and groundwater.

Several problems related to using waste organic materials and fixed analysis chemical fertilizers can be detected in Table 9–2C. In the last portion of the table, it can be seen that when 8,000 pounds (3,635 kg) of farmyard manure is applied per acre to meet the crop removal of nitrogen, a surplus of phosphorus and potassium is created. The crop only requires that 5,600 pounds (3,635 kg) and 6,600 pounds (3,000 kg) of farmyard manure to meet the total crop removal of phosphorus and potassium, respectively. In a similar fashion, if 18,857 pounds (8,571 kg) of night soil is applied to meet requirements for potassium, there is a surplus of nitrogen and phosphorus that only needed 6,667 pounds (3,030 kg) and 3,590 pounds (1,632 kg) of night soil, respectively, to provide for crop removal.

Similar problems also result with fixed formula chemical fertilizers. If 471 pounds (214 kg) of 17–17–17 fertilizer is applied to meet the nitrogen requirements, the potassium requirement is very nearly matched, but more than twice the amount of phosphorus, which only needed 187 pounds (85 kg) of 17–17–17 fertilizer, is added. The excess nutrient applied to any one crop is a potential environmental pollution problem, and in the case of purchased chemical fertilizer it represents an expenditure of money with no immediate return in crop yield.

There are several fixed formula chemical fertilizers available that allow farmers and gardeners to select a formulation that most closely matches the requirements of the crop to be planted. Also, modern management technology allows farmers to purchase and apply custom blended fertilizer to more nearly match the exact ratio of nutrients needed by the crop. In this example 400 pounds (182 kg) of a 20–8–20 blended fertilizer very nearly matches the crop removal of nitrogen, phosphorus, and potassium. Custom blended fertilizer can be applied at variable rates and formulations to accommodate spatial differences in nutrient content identified by soil test analyses among various soils within the same field. This practice is known as *precision farming*.

Comparisons of the amounts of material required illustrate that the most serious limitation to the use of waste products and manures is the cost and/or the labor involved in transporting the material to the field where the crop is grown. In almost every scenario calculated in Table 9–2, more than ten times as much weight of a waste product, as compared to a concentrated chemical fertilizer, is required to meet the expected crop needs. In societies lacking infrastructure, the method of transport is human labor. In societies with highways and trucks, the distances from the waste products to food crop fields are often great and the cost of transportation excessive.

In small villages in tropical jungles, the labor requirements for transporting waste materials to fertilize crops is clearly evident in the land use patterns around the villages. Ruthenberg (1971) diagrammed food-growing practices as a pattern of concentric circles around small villages in tropical jungles. Continuous gardens are grown near the houses at the center of the circles. The next distant circle is intensively used for semipermanent cultivation of food crops. At somewhat greater distances short rotation-shifting cultivation is practiced, and finally on more distant land food crops are grown by slash-and-burn methodologies after periods of fallow. The gardens and nearby areas used for continuous cultivation have their fertility renewed by applications of night soil and other household wastes. The slightly more distant semipermanent areas received occasional applications of manures, and the more distant land relies on the natural regrowth of vegetation that is burned at intervals of several years to provide ash to fertilize food crops. The amount of labor involved in carrying waste materials is clearly a limiting factor in their use as fertilizer.

As urban populations increase and food sources become more remote, the labor and transportation costs of transporting waste materials increases. Organic waste tends to be produced every day. Adverse weather and seasonal conditions of wetness or cold often curtail access to spread waste material on the crop fields. It is certainly not desirable to spread waste material that may contain pathogenic organisms on a crop that is about to be harvested. A waste storage capability is required. Storing waste introduces problems of odor, insect, and pathogen control.

In most mechanized societies, concentrated and chemically processed fertilizers are much more economical than distributing waste materials for fertilizing food crops. For example, the application of 20 pounds (9 kg) of phosphorous requires only about 100 pounds (45 kg) of inorganic triple super phosphate (20% phosphorus) fertilizer, whereas 5,000 pounds (2,273 kg) of night soil would be required to supply the same amount of phosphorus. Although there may be no cost for the essential elements in waste products, their application to the site of food production is usually very expensive.

THE RIGHT TIME TO FERTILIZE

Timing, the inescapable ingredient in obtaining maximum efficiency from fertilizing material, is an often unseen but critical concern in using organic waste materials to fertilize soil used for food crop production. It is best to fertilize cropland immediately prior to the intense growing period of the food crop.

Organic materials must be decomposed to inorganic ions before plants can utilize the essential nutrient elements (Figure 5–1). The rate at which organic materials are decomposed by microbes after their application to the soil depends on many factors, such as type of organic material, temperature, and moisture availability. Grain crops, most notably corn, have a very high demand for nutrients, especially nitrogen, during a very short period of time while they are rapidly growing. To meet this demand, readily soluble inorganic nitrogen, often as ammonium nitrate (NH_4NO_3) or anhydrous ammonia (NH_3), are applied below the soil surface adjacent to rows of corn plants 20 to 30 days after the corn is planted. The inorganic nitrogen ions are rapidly dissolved into the soil water and available to the rapidly growing corn, assuring maximum utilization of the fertilizer and minimum potential for nitrate leaching into the groundwater. This practice is commonly known as *side dressing*.

To assure adequate amounts of plant-available nitrogen during peak periods of crop demand, excessive amounts of organic material must be applied far enough in advance of the peak demand that microbes will mineralize sufficient organic nitrogen, making it available as inorganic NH_4^+ and NO_3^- ions for the crop. Microbial decomposition of organic waste materials is a rather slow process, and several cycles of decomposition are required for complete release of nitrogen and the other essential elements. Microbes also continue to decompose the organic material after the crop has reduced its uptake of nitrogen, and the nitrate ions (NO_3^-) mineralized from the organic materials after crop growth has slowed are subject to leaching (Pang and Letey, 2000). In temperate latitudes where winter temperatures prevent continuous crop growth, the nitrate (NO_3^-) formed from the decomposition of organic materials after the crop is harvested remains in the soil water into the winter season. Recall from Chapter 2 that it is during the winter and spring that most leaching and groundwater recharge takes place in temperate latitudes with humid climates. Excess nitrate (NO_3^-) in the soil during the winter and spring is very susceptible to leaching and a potential threat to the groundwater. Soluble inorganic nitrogen fertilizer applied just before maximum crop growth and immediately available to the crop plants poses a lesser threat to the groundwater.

It is often tempting but incorrect to consider the return of straw or other parts of a crop plant not removed with the harvest of the grain as a source of nutrients. Although the uneatable parts of food crops such as straw do contain essential nutrient elements, these elements have been obtained from the soil where that crop is grown. Although the direct return of as much of the unusable plant material (usually the straw of stem tissue) as possible is desirable, these practices *add* no essential elements originally derived from soil minerals because the essential elements contained in the straw were obtained from that site.

The harvest of many food crops requires that grain be separated from straw. It is necessary to dry the entire plant to effect good separation. If weather conditions cooperate, the separation of grain and straw can be done in the field where the crop is grown, thus immediately returning the nutrients in the straw to the soil. Although this is commonly done with highly mechanized equipment like combines, societies limited to hand labor usually gather the entire aboveground portions of the grain plant and protect them from the weather so they dry prior to grain separation. The straw is often used for animal food or burned for heat and cooking at the house site or village. Even when returned immediately to the field, most straw has such a high carbon-to-nitrogen ratio that it may suppress nitrogen uptake by the next crop (Table 4–3). The addition of organic residues with carbon-to-nitrogen ratios greater than 32:1 actually decreases the amount of available nitrogen in the soil as microbes seek nitrogen to achieve a 10:1 carbon-to-nitrogen ratio in their cells. To avoid this, straw deposited directly in the field by modern combines at harvest is often burned to remove carbon, although some nitrogen is lost in the fire.

Because elements can neither be created nor destroyed, those elements we eat must be replaced in the soil, at the site on the land from which they came, if food production from that land is to be sustained. There is cost to replenish the nutrients harvested in food. Only if farmers are assured by the surrounding society that fertilizer costs can be recovered via reliable marketing of the products will they fertilize and continue to cultivate the same area and produce food in excess of the family needs. Where the surrounding society does not support the cost of enriching the soil by application of essential elements from external sources, farmers adopt subsistent production practices to feed their immediate families and provide little food for other segments of the society. There is no free lunch.

FARMING CHANGES ON SLIGHTLY FERTILE LAND

Perhaps the most extreme examples of food production schemes on slightly fertile land are those used by the nomadic societies. In such societies the population density is low per unit of land area. Human food is either directly obtained from eatable plants in the ecosystem or indirectly by using animals to graze the native plants and then consuming the animal products of meat and milk. Human energy is used to move the human individual, or animal, to the site of food production rather than moving the food to the site of consumption. Waste products are dispersed and in a sense totally recycled. Historically, nomadic societies have existed in most parts of the world and persist in some places today.

Where societies seek to establish more permanent residence, some form of slash-and-burn farming is usually practiced. In many tropical forests this form of a low input, low output system of securing food from land with moderate to low contents of essential elements is used today (see Chapter 8, "Warm, Watered, and Slightly Fertile"). Slash-and-burn systems have been historically practiced in most parts of the world. Slash-and-burn techniques were especially prevalent in the southeastern portion of the United States as European settlers attempted to farm the forested but fertility-limited Ultisols (see the appendix) that are the most prevalent soils in the region. The words of Professor Mitchell in 1822 (Mitchell, 1822) describe how farming was conducted by the first European settlers in North Carolina and vividly outlines experiences common to all slash-and-burn systems:

> When our ancestors landed on these shores, they had for ages been covered with a continued forest, the trees of which, as they decayed and fell, had deposited on the earth a rich bed of vegetable matters, which was ready to furnish the most abundant nourishment to any seed that might be committed to the ground. The first settlers, therefore, had nothing to do but to select the most promising spots, clear away the timber, and loosen the soil, so that the vegetables to be grown could strike their roots into it. As the fertility which they had at first found was, in the course of a few years, exhausted, it became necessary, either to provide the means of renewing it, or deforest another tract and bring it under cultivation. As it was found that the latter could be done at the least expense of time and labor, it was perfectly natural that the exhausted land should be thrown out, and fresh ground brought under tillage.
>
> This process has been going on till most of the tracts whose situation and soil were most favorable to agriculture, have been converted into old fields and in our search after fresh ground to open, we are driven to such inferior ridge-land as our ancestors would have passed by as not worth cultivating. It is useless to complain of the course which our planters have pursued—they have pursued their own interest—and pursued it in the main with discretion and judgment. It were perfectly absurd to expect them to attempt to improve their lands by the application of manures so long as they could obtain, at less expense, the use of that great store of vegetable matter with which nature had for many centuries been covering our country. It is not to be expected that a man will raise a hundred barrels of corn in a way which we may point out to him as best, at an expense of three hundred dollars, when his past experience informs him that he can produce it in his own way for two hundred. But, in process of time, as this system goes on, the planter will look down from the barren ridges he is tilling, upon the ground from which his fathers reaped their rich harvest, but which are now desolate and abandoned, and enquire whether he cannot restore to them their ancient fertility, at a less expense than he can cultivate those lands of inferior quality, with which he is now engaged. Till he is driven by necessity to make this enquiry, we can hardly hope that agriculture will be studied as a science. The planter will not give us a patient hearing when we talk to him about manures.

What Mitchell did not describe, and could not see while slash-and-burn practices were actively used, was the fact that when the abandoned fields were undisturbed for several years, a new cycle of slow-growing native tree vegetation would again populate the abandoned lands. The first few years after abandonment, erosion on sloping land would be rapid, but grass and weeds would soon establish and slow the rate of erosion (Figure 9–5). With time, and reseeding by wind and birds, tree species native to the area would reestablish on the land. In that part of North Carolina described by

FIGURE 9–5   Photo of an abandoned slash-and-burn field in western North Carolina, USA, in 1902.

Mitchell, the first trees to vegetate the land would be loblolly pine (*Pinus teada*). They would grow to maturity in about 75 years during which time hardwoods of oak, hickory, popular, and dogwood would establish and become the climax vegetation. During this period of forest succession, minerals within the soil would slowly weather and yield a small but adequate quantity of essential elements in plant-available forms for the slow-growing trees. From annual leaf fall and the ultimate death of the trees, these elements would accumulate as organic compounds in the surface layers of the soil now protected from rapid erosion by the vegetation. If settlers were to again invade the area after several years of natural regrowth, the scenario described by Mitchell could be repeated.

The similarity of historical slash-and-burn cultivation in the southern United States with present slash-and-burn farming in many parts of the tropics can be seen by comparing Figure 9–5 with Figure 9–6.

However, many technical and social changes took place in the southern United States during the early part of the 20th century. Many of the early European settlers moved west to farm the more fertile soils of the Midwest of the United States. Farmers that remained on the limited fertility soils of the southeast concentrated on the production of high-value crops and became dependent on chemical fertilizer to replenish elements harvested in their crops.

FIGURE 9–6 Photo of slash-and-burn farming on the eastern slopes of the Andes in Peru, SA, in 1985.

It is informative to review the late-19th-century publications of various Agricultural Experiment Stations funded by the federal Hatch Act. The Hatch Act was initiated after the Civil War and provided funds for agricultural research at state colleges known as land grant colleges. These colleges developed research, teaching, and agricultural extension services as major components of their faculties, and because they were federally funded wrote annual reports. With few exceptions, the federal research funds in the southern states were used to study soil fertility and fertilizer; federal funds at the midwestern land grant colleges were used primarily for animal research.

Much, if not almost all, of the most erosion-susceptible land in the southeastern United States that had been cleared in slash-and-burn farming was abandoned for further cultivation after World War II. Although some small fields on steep slopes were fertilized for the growing of labor-intensive high-value crops like tobacco, most were allowed to return to natural forest. Engineered drainage systems were constructed in the bottomlands where continuous cropping could be conducted with annual applications of fertilizer and lime. Another factor that contributed to the abandonment of the steep erosive-prone land was the introduction of tractors and larger cultivation and harvesting equipment. A farmer simply could not safely operate large equipment on the steep slopes that had previously been cultivated with hand hoes or small mule-drawn equipment. The evidence of the old slash-and-burn fields on steep slopes in the mountains of North Carolina can be seen by a careful examination of the rectangular pattern of vegetation on the hillsides in Figure 9–7. Also note that engineered drainage has replaced the meandering streams in the bottomland seen in the photo.

FIGURE 9–7   Photo of a mountainous area once used for slash-and-burn farming in North Carolina, USA, as it was in 1982. Note the rectangular pattern of regrowth vegetation outlining abandoned fields on the hillside and the constructed drainage canals in the pastured valley bottom.

FARMING CHANGES ON FERTILE LANDS

There was a rapid influx of European farmers into the more chemically fertile and well-watered land of the Midwest in the United States after about 1850. Unlike experiences on the slightly fertile Ultisols in the southeastern United States, crop yields did not decline rapidly after the first few crops were harvested on the fertile Alfisols and Mollisols (see the appendix) in the Midwest. The same land could be planted with corn, oats, or wheat every year with the expectation of a reasonable yield.

In 1876 a well-documented long-term corn-growing experiment was established on the University of Illinois campus, a land grant institution, and continued each succeeding year for more than a century (Odell et al., 1982). When we examine the yields in plots of continuous corn during the first ten years of available data (1888–1897), we find the yearly average yield to be 40.9 bushels (56 Lb Bu^{-1}) of corn per acre (2,565 kg ha^{-1}) (Table 9–3A). No manure, lime, or fertilizer was applied to any of the experimental plots during this period. Starting in 1904, manure, lime, and rock phosphates were added annually to a portion of the previously untreated plots. During the next ten years (1905–1914), yields averaged 40 Bu Ac^{-1} (2,509 kg ha^{-1}) on those plots, and the average yield in the plots receiving no manure, lime or phosphate dropped to 26.1 Bu Ac^{-1} (1,637 kg ha^{-1})(Table 9–3B). Clearly even these naturally fertile soils were not able to sustain their original-yield levels without an external supply of essential elements.

TABLE 9–3A Morrow Plot Continuous Corn Yields, 1888–1897

Year	No Fertilizer	Manure, Lime, Rock Phosphate	High Input
		Bu Ac⁻¹	
1888	54.3	ND	ND
1889	43.2	ND	ND
1890	48.7	ND	ND
1891	28.6	ND	ND
1892	33.1	ND	ND
1893	21.7	ND	ND
1894	34.8	ND	ND
1895	42.2	ND	ND
1896	62.3	ND	ND
1897	40.1	ND	ND
1888–1897 Average	40.9	ND	ND

Note: A bushel of corn weighs 56 lb (Bu Ac^{-1} × 56) × 1.12 = kg ha^{-1}. ND, Not done.
Source: Odell et al., 1982.

The practice of adding complete chemical fertilizers, in amounts recommended by soil test analyses, was started in 1967 on a portion of the Morrow plots and termed *high input* in Table 9–3C. During the 1972 to 1981 decade, an average yield of 131.9 Bu Ac^{-1} (8,273 kg ha^{-1}) was reported on the high-input plots compared to 47.7 and 80 Bu Ac^{-1} (2,992 and 5,018 kg ha^{-1}) in the untreated and manured plots, respectively. Note that one "disastrous" year, of extreme drought, significantly lowered the ten-year average that is 143 Bu Ac^{-1} (8,969 kg ha^{-1}) for the other nine years. We should be quick to recognize that corn production, even on some of the best soils in the world, is not guaranteed and

TABLE 9–3B Morrow Plot Continuous Corn Yields, 1905–1914

Year	No Fertilizer	Manure, Lime, Rock Phosphate	High Input
		Bu Ac⁻¹	
1905	24.8	26.8	ND
1906	27.1	32.5	ND
1907	29.0	40.8	ND
1908	13.4	24.8	ND
1909	26.6	30.4	ND
1910	35.9	48.9	ND
1911	21.9	29.0	ND
1912	43.2	64.4	ND
1913	19.4	32.4	ND
1914	31.6	37.2	ND
1905–1914 Average	26.1	40.0	ND

Note: A bushel of corn weighs 56 lb (Bu Ac^{-1} × 56) × 1.12 = kg ha^{-1}. ND, Not done.
Source: Odell et al., 1982.

TABLE 9–3C Morrow Plot Continuous Corn Yields, 1972–1981

Year	No Fertilizer	Manure, Lime Rock Phosphate	High Input
		Bu Ac^{-1}	
1972	57.9	101.9	159.10
1973	44.0	78.1	148.6
1974	39.4	72.6	105.2
1975	48.8	76.3	161.3
1976	49.6	82.8	135.5
1977	51.1	87.7	113.0
1978	58.3	98.4	176.8
1979	35.2	63.0	115.9
1980	46.3	56.8	31.1
1981	47.0	82.5	172.3
1972–1981 Average	47.7	80.0	131.9

Note: A bushel of corn weighs 56 lb (Bu Ac^{-1} × 56) × 1.12 = kg ha^{-1}. ND, Not done.
Source: Odell et al., 1982.

fluctuates in accord with weather conditions during the growing season. Also, note that the yield in high-input plots was more severely reduced than in the less fertilized plots by the drought in 1980. This apparent susceptibility of a well-fertilized crop to adverse weather conditions is rare and often related to weather events at a particular stage of physiological development. Corn is most vulnerable to drought at the time of pollination that may have been somewhat different between the compared plots.

Many genetic changes have been made in the type of corn grown and planting density over the time the Morrow plots were studied. The introduction of hybrid varieties may in part account for the yield increase in both the unfertilized and manured plots. The continued production of corn at a level of about 45 Bu Ac^{-1} (2,822 kg ha^{-1}) for nearly 100 years in plots receiving no manure or fertilizer is remarkable and does not occur in most soils of the world. The soils at the site of the Morrow plots are Mollisols (see the appendix), some of the most naturally fertile soils in the world. Although not reported for the Morrow plot site, it is well known that both the topsoil and subsoil of Mollisols developed from glacial till and loess and contain very high contents of phosphorus and potassium as represented by the data from the USA Mollisol cited in Table 5–1.

Nitrogen is probably the most limiting essential element for corn production in unfertilized Mollisols. It is reasonable to estimate that the Morrow plots annually receive about 20 pounds of nitrogen per acre (22 kg N ha^{-1}) from rainfall and non-symbiotic nitrogen-fixing bacteria and perhaps another 20 to 25 pounds of nitrogen per acre (22–28 kg N ha^{-1}) from the decomposition of soil organic matter. That amount of nitrogen would provide the nitrogen necessary for a 45 Bu Ac^{-1} (2,822 kg ha^{-1}) yield. Odell et al. (1982) reported that soil organic matter contents continued to decline during the 100 years of continuous corn production, indicating that a steady release of nutrients from the organic pool was taking place.

The inescapable conclusion from the Morrow plot data is that the natural ability of even the most naturally fertile soils in the world to produce high yields of food crops is limited by a natural supply of essential elements. When only the quantities of essential chemical elements available from rain and the normal capture of nitrogen by nitrogen-fixing microbes are supplied, yields remain low. A more recent report of all corn yields in the United States shows that average yields have increased from approximately 74 Bu Ac^{-1} in 1974 to 149.4 Bu Ac^{-1} in 2004 (USDA-National Agricultural Statistics Service, 2004). Fertilizer is now applied to nearly all corn and other food grain crops in the United States. With nationwide corn yields now being more than three times greater than those obtained on the most naturally fertile but unfertilized plots at the University of Illinois, it is easily seen why considerably less land is now needed to feed a growing U.S. population (Figure 9–1).

FEEDING CONCENTRATIONS OF PEOPLE

Throughout the history of civilization, humans have concentrated in cities and sought to intensify food crop production on rural land. The wheel was a major tool allowing people to construct carts and wagons that helped carry food in greater quantities to urban centers. Perhaps the most primitive agricultural practice was irrigation, which permitted people to protect their crops against some of the uncertainties of weather. Vessels for carrying water are found among the ruins of many ancient peoples. Although such vessels were used to carry water for human consumption, many ancient drawings portray their use in supplying water to crop plants.

When there is insufficient water in the soil to supply the immediate plant needs, the plants visibly respond by wilting. Recovery from a wilted condition is often rapid if water is applied to the soil around the plant roots. Humans recognized plant needs for water early in the development of civilization. Irrigation from a reliable water source could provide continuous fresh food in warm tropical areas. In most early civilizations, irrigation water was obtained by construction of elaborate canals through which water from rivers and streams could be diverted to fields devoted to the growth of food crops. In other areas, water was lifted from wells dug to the depth of reliable water tables. The unseen component of successful irrigation in early centers of civilization was the chemical renewal provided by sedimentation from floodwater on the nearly level bottomland and floodplain areas that were irrigated.

As human and animal power were replaced with mechanical and fuel-based sources of energy, irrigation practices became more sophisticated. Pumps that could lift water from greater well depths were installed. Dams were constructed to hold water for irrigation during rainless seasons. The construction of the Aswan Dam in southern Egypt made it possible to grow three crops per year on much of the land in the Nile river valley that before could grow only one crop each year following the annual flooding. However, controlling the natural flooding stopped the input of essential elements derived from the erosion of soils in the watershed and suspended in the floodwaters. With yearlong irrigation available, two or three crops per year could be grown each year, creating a greater demand for nutrients from the soil. Less time was available for the capture of nitrogen from the air and release of phosphorus, potassium, nitrogen, and calcium from organic matter decomposition and weathering of soil minerals. Fertilization became necessary in the ancient fields.

As recorded in the Bible (Luke 13:6–9), the early cultivators knew that crop growth could often be improved when animal manure was added to the soil. It was believed that humus (mainly carbon) came from the soil to be recreated into plant parts by the next generation of plants. Andrews (1954) pointed out that by about 1700 people farming fertile land maintained or increased crop yields by using animal manures, green manures, crop rotations, and lime, but these practices did not supply sufficient plant nutrients to maintain crop yields on less fertile soils. As seen in Figure 9–8, before about 1730, wheat (60 Lb Bu^{-1}) yields in England were about 8 Bu Ac^{-1} (538 kg ha^{-1}). By 1840 farmers applied manure and used crop rotations that included one growing season in which legumes were grown but not harvested and the entire plant was mixed into the soil to provide nitrogen for the next crop. With these practices average wheat yields increased to about 20 Bu Ac^{-1} (1,344 kg ha^{-1}).

A major discovery that was to affect the entire course of food production in the world emerged during the middle of the 19th century. In 1840 Justus von Liebig declared that the carbon plants needed came from the air, whereas it was necessary for the plant to obtain phosphorus and alkali salts (potassium, calcium, etc.) from the soil (Russell, 1952). In 1842 J. B. Lawes patented a method for producing superphosphate and started the Lawes Chemical Company in England for its manufacture along with other "manures" (Anonymous, 1977). By 1855 it was known that plant growth increased when these elements were added to soil as "artificial manures" (mineral salts of essential elements now in chemical fertilizers) rather than real animal manure. It was recognized that if the physical conditions in a soil were favorable for plant growth, the value of animal or other organic manure was not organic carbon but "depends only on the amount of nutrients it can supply in simple form to the crop" (Russell, 1952, p. 25). We now know that 'simple form' to be inorganic ions in the available pool (Figure 5–1).

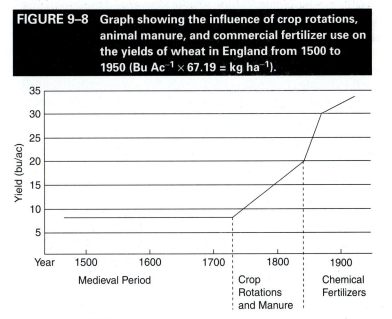

FIGURE 9–8   Graph showing the influence of crop rotations, animal manure, and commercial fertilizer use on the yields of wheat in England from 1500 to 1950 (Bu Ac^{-1} × 67.19 = kg ha^{-1}).

Source: Andrews, 1954.

By 1870 the use of inorganic chemical fertilizers sharply increased average wheat yields in England to 30 Bu Ac^{-1} (2,016 kg ha^{-1}). Average yields continued to increase in the early part of the 1900s, and by the end of the century, average wheat yield in England was over 90 Bu Ac^{-1} (6,048 kg ha^{-1}) (Greenland, 1997).

Much research followed the recognition that the carbon in plants came from the air and inorganic forms of most other essential elements came from the soil. In attempts to identify the specific minerals that were most efficient in transferring the needed elements from the soil to the plant, numerous studies were made of microbes and chemicals that made the essential elements in the soil minerals more available to the plant. Various acids and other "solubilizers" were added to the soil to increase the availability of essential elements in soil minerals, but none proved able to provide the plant with a sufficient supply (Russell, 1952). It became clear that even solubilizing the total amount of essential elements present in soil was not sufficient to sustain food crop production. Supplying quantities of essential elements via inorganic chemical fertilizers from external sources became the norm for intensive food crop production.

Maintaining the chemical supply of elements removed in crop harvest with inorganic forms of the elements proved to be a challenging task. It was difficult to formulate inorganic fertilizers that would compete with "good old animal manure." After all, animal manure contains all the elements necessary for plant growth, in proportions approximating those required by plants. Inorganic fertilizer formulations often lacked some of the elements required in small amounts, known as trace elements, such as boron, copper, zinc, iron, and sulfur. The law of the minimum (Figure 4–1) must be obeyed. As trace element problems became better understood by agricultural scientists, manufacturers of inorganic fertilizer were able to solve most problems that plagued the early development of chemical fertilizers.

Although inorganic fertilizers were known and used for many years prior to the 1940s, it was not then that a substantial increase in chemical fertilizer use took place in the United States. As seen in Figure 9–1, before 1940 the amount of land used to cultivate food crops for consumption within the United States increased as the population increased. A number of factors influenced the dramatic reversal of that trend during the decades that followed. Energy became abundantly available to power the conversion of N_2 from the air into forms of nitrogen easily transported and applied to the soil. Large chemical plants were built that could economically concentrate phosphorus mined from mineral rock phosphate deposits into soluble compounds that could be easily transported and applied to crop fields (van Straaten, 2002). Similar processes were perfected for concentrating potassium from potassium-rich rocks. Reliable standards were established for crushing limestone into agricultural lime for correcting acid soil conditions as well as supplying calcium and magnesium. Both the improvement of rural roads and trucks to transport food crops from the fields to the expanding export markets of Europe and Asia after World War II gave financial incentives to farmers in the United States to increase their production. Quantum leaps in fast and powerful farm machinery made it possible for farmers to carry out the weather-dictated tasks of plowing, planting, weeding, and harvesting of many more acres per day than had previously been possible with small tractors or animal or human power. As tractors replaced horses and mules, the supply of manure available on farms was reduced and the amount of grain previously fed to the horses and mules was available for human consumption and feeding animals that produced meat and milk.

Concentrated inorganic fertilizers fit well into the pattern of farming that developed in the United States and Western Europe. Farmers that owned land well suited to crop production expanded their holdings of such land and concentrated their skills on specific types of crop production. Human talent of farmers became more specialized, and each operation became more efficient. Animal production that depended on a source of grain was concentrated onto small areas, feedlots, where animal care and feeding could be more efficient. But animal waste in feedlots was now distant from the fields where the animal food was grown, and animal wastes became localized problems.

Reliable transportation was perhaps the most influential component of this agricultural transformation—not only the physical transportation of farm products but a reliable flow of money and credit from one specialist to another throughout the complex agricultural food chain of farmer, food processor, supermarket, banker, seed producer, fertilizer producer, and consumer. A multitude of industries such as machinery manufactures, fuel suppliers, refrigeration specialists, and so on, play a vital part in this loop that makes the variety and quality of food we presently enjoy possible. We must also recognize that such a complicated flow of money and material cannot be developed or maintained without social and political stability. As individuals specialized, the size of successful farming operations increased. Several farms were consolidated under ownership of one individual. Crop production became concentrated on land most suited by reason of topography and soil conditions of temperature and rainfall. Farmers concentrated on only a few crops, thus reducing the need for several kinds of specialized farm machines. Small farm implement dealerships, hardware supply stores, and repair services were consolidated. Rapid communication and transportation could service the needs of the farmer. Small banks were taken over by large financial conglomerates that serviced the entire agricultural infrastructure from farm to market. Figure 9–9 makes evident what this scenario has meant to the average consumer in the United States.

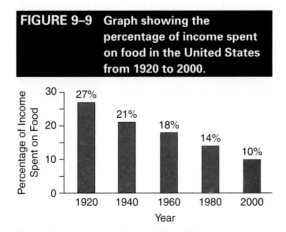

FIGURE 9–9   Graph showing the percentage of income spent on food in the United States from 1920 to 2000.

Source: National Association of State Universities and Land-Grant Colleges, 1997.

CROP FERTILIZER NOT SOIL FERTILIZER

Although fertilizer is used to alter the chemical composition of the soil, it is the harvest of nutrients in the crop plant that determines the need for fertilizer in sustained crop production. Table 9–4 summarizes one specific example of fertilizer use over time.

Soil survey reports of Wilson County, North Carolina, in 1925 and 1983 contain the yields of corn expected on Norfolk (Ultisols; see the appendix) soils during the dates indicated. Similar reports from Cedar County, Iowa, in 1919 and 1979 contain expected corn yields for Tama (Mollisols) soils during each of the dates indicated. Tama soils are considered some of the most naturally fertile soils in the world, similar to the Flannagan soils at the Morrow Plots at the University of Illinois already discussed. Norfolk soils are representative of some of the naturally least fertile soils in the United States, although certainly not the least naturally fertile soils in the world. Both Tama and Norfolk soils are some of the most desirable agricultural soils present in the respective states where they are present. Yield estimates in the soil survey reports reflect yields that farmers, using good management practices, obtain on the respective kinds of named soils.

Note that on both the Norfolk and Tama soils, corn yields in about 1980 were approximately triple the yields in about 1920. The Tama soils are clearly better for corn production than the Norfolk soils in both periods of time. This may be due to several factors: longer day length, cooler nighttime temperatures, and better subsoil fertility in Tama soils in Iowa than in Norfolk soils in North Carolina. Also, note that farmer yields on nonfertilized Tama soils in 1919 were nearly the same as the yields maintained for nearly 100 years on the similar Flannagan soils at the Morrow Plots in Illinois when no fertilizer was added and continuous corn crops were grown (Table 9–3C).

On the Norfolk soils in North Carolina, farmers were already using small amounts of fertilizer in the 1920s but failed to obtain corn yields as high as yields without fertilizer in Iowa. There are no records to document fully the history of cropping in either area prior to the 1920s, but it is probable that farming had been conducted in Iowa for about 50 years and for a longer period of time in North Carolina. Other studies indicate that without fertilizer or manure, it was not possible to obtain any reasonable corn yield on soils like Norfolk for more than a few years once cropping started (see the Mitchell 1822 quote earlier in this chapter). The huge increase in fertilization rates between the

TABLE 9–4	Historical Comparison of Average Farmer Fertilization Rates on a Naturally Fertile and a Naturally Infertile Soil in the United States[*]				
	Element	*Norfolk Soils (NC)*		*Tama Soils (IA)*	
Year		*1925*	*1983*	*1919*	*1979*
Corn Yield (Bu Ac^{-1})		32	110	42	130
Fertilizer Rate (Lb Ac^{-1})[†]	N	32–47	120–158	0	150–180
	P	3–5	18	0	30–48
	K	5–10	67	0	67–99

[*]Original data from Soil Survey Reports and Unpublished Agricultural Extension Service records.
[†]Lb Ac^{-1} × 1.12 = kg ha^{-1}.

1920s and the 1980s, both in Iowa and North Carolina, correspond to the yield increases in both areas.

When the rates of fertilizer use on both the Tama and Norfolk soils about 1980 are compared, the records show that more fertilizer was used on the more naturally fertile Tama soils than on the naturally infertile Norfolk soils. The reason for this is reflected in a comparison of the yields obtained. When compared to the elemental requirements of corn (Table 5–4), it is clear that the standard farmer practice in both Iowa and North Carolina by 1980 was to apply only slightly more fertilizer than needed to supply average yields expected. A subjective reason for the slight overapplication of fertilizer is that the yield data are for average years, and farmers plant with optimism that the coming year will be better than average. Therefore they want to make sure their crop yield will not be limited by lack of nutrients if the season is exceptionally good. Also, there is always a small amount of nutrient loss due to fixation and leaching so a 100% return of nutrients applied as fertilizer is not possible. It is clear that although farmers apply fertilizer to the soil, the amount of fertilizer they apply is determined by the expected crop yield irrespective of the natural fertility of the soil.

NUTRITIONAL QUALITY OF PLANTS

Increased nutrient element concentration in the soil also results in increased nutrient value of the plants grown on that soil (Grunes and Allaway, 1985). This enrichment benefits animals and humans consuming the plants as food. As seen in Figure 9–10, unfertilized oat grain barely contains enough phosphorus to be a satisfactory food source for dairy cattle. Even alfalfa, grown on phosphorus-deficient soils does not contain enough phosphorus for dairy cattle until about 22 Lb Ac^{-1} (25 kg ha^{-1}) of phosphorus

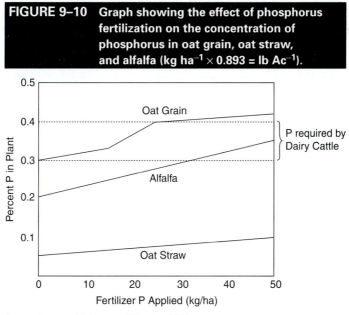

FIGURE 9–10 Graph showing the effect of phosphorus fertilization on the concentration of phosphorus in oat grain, oat straw, and alfalfa (kg ha^{-1} × 0.893 = lb Ac^{-1}).

Source: Larson, Nelson, and Hunter, 1952.

fertilizer was added to the soil. Oat straw does not contain enough phosphorus to be a satisfactory food for dairy cattle regardless of the amount added to the soil.

Most plants are able to survive and grow to maturity within a range of nutrient element availability in the soil (see Figure 4–2). Animals often have a more narrow range of tolerance to the nutrient content of their food. Thus the same species of plant may in one case be a nutritious food when grown on a fertile soil but provide inadequate nourishment when grown on a nutrient-deficient soil. In certain cases toxicities can result from excesses of specific elements, but these cases are rare, localized, or result from abnormal weather conditions that favor rapid increases of certain elements in the available pool.

All plants are not equal in either demand for, or tolerance to, available nutrient concentrations in soil. In natural ecosystems, a limited number of plant species tend to populate specific kinds of soils in response to the elemental composition of the available pool. Even trees of the same species growing on soils with low concentrations of essential elements have lower contents of those elements in their leaves than trees growing on soils with higher quantities of the same element (Table 5–3). Thus the amounts of essential elements released when equal amounts of forest biomass are burned in slash-and-burn systems are less when that biomass has grown on infertile soil than when grown on fertile soils.

In ecosystems managed for crop production, the elemental composition of the available pool is managed to optimize the growth of the desired crop by applying fertilizer and adjusting soil pH values with lime. The same soil will have different properties when managed for crop production than if left under unmanaged natural conditions. Analogous comparisons could be made to the morphology of a bush that is trimmed as part of a sculptured hedgerow to the same bush allowed to grow freely with no trimming or the appearance of a feral dog compared to a well-groomed dog in a dog show.

To attain high yields of food crops, the chemical composition of a soil must be carefully managed in accordance with the requirements of the plants growing on that soil. Invariably, this requires increasing the amount of essential elements in the available pool during the time fast-growing food crops are being grown even on the most naturally fertile soils. Because different plants have different chemical compositions, it is necessary that the relative proportions of each essential element in the soil be determined so as not to introduce an access of any given element. Fertilizer formulization with respect of the elements contained should conform to the requirements of each type of crop grown. Within general guidelines, various fertilizer formulizations are known as corn fertilizer, lawn fertilizer, garden fertilizer, and so on.

Every discussion of supplying essential elements via either organic residues or chemical fertilizers *must include the need to maintain a suitable pH value* in the soil to assure availability of those elements. Most naturally acid soils require applications of lime before fertilizer is effective. Continued applications of lime are required to offset the natural acidifying effect that growing plants have as they exude H^+ from their roots. Deficiencies of iron, copper, zinc, boron, and manganese may result if the soil pH is increased above 7.5. Human vigilance of the chemical environment within the soil is paramount to sustaining high levels of food crop production.

Of particular concern in management of soils for sustained crop production is application of fertilizing materials of unknown or variable composition. In most countries the composition of manufactured commercial fertilizers is carefully controlled and identified to the user. This is not easily done with manure and sludge wastes in which the content of nutrient elements is variable. Some waste materials may contain heavy metals such as arsenic, cadmium, chromium, copper, lead, mercury, molybdenum, nickel, selenium, and zinc that can raise concentrations of the elements in the soil to toxic or lethal levels. Once excessive amounts of heavy metals are introduced into a soil, they are extremely difficult to remove. Plants grown on soil with excessive concentrations of heavy metals may accumulate the heavy metals and cause physiological problems when consumed by animals and humans. To reduce excessive heavy metal concentrations in plants growing on soils containing excessive amounts of heavy metals, it is critical to control the pH value of the soil. The availability of all heavy metals is not affected in the same way by pH. At higher pH values, the plant availability of most heavy metals is reduced, but the plant availability of arsenic, molybdenum, and selenium is increased. It is prudent to be aware of heavy metal content in any material added to the soil.

HOW MUCH CROPLAND?

Farmers receive income from the total amount of crops they sell. It is almost always of benefit to them to generate the highest possible yield per area of land because of the fixed costs of taxes, interest on market value of the land, and the investments in fuel and labor involved in cultivating and harvesting each unit of land.

High yield per unit of land also has profound implications for national and world land use. Compare corn yields of about 40 bushels per acre (2,509 kg ha^{-1}) on the very naturally fertile Flannagan and Tama soils without the use of fertilizer with the 2004 national average corn yield average of 149 bushels per acre (9,342 kg ha^{-1}) (USDA-National Agricultural Statistics Service, 2004). Yields of other grain crops have similar yield increases with fertilization. One can conclude that if only unfertilized crop yields were attained today, more than three times as many acres of land would now be needed to produce the quantity of food the present population of the United States consumes. To visualize what this would mean to the land use in the United States, go to Figure 9–1 and extend the "Acreage harvested for U.S. consumption" line from 1910 to 1940 forward to the year 1990 and beyond. From that you can estimate that by 1990 somewhat over 400 million acres (162 million ha) of cropland would have to be harvested for U.S. consumption. This is approximately 100 million acres (40 million ha) more than the United States cultivated to feed its own population and exported to the world markets in 1990. Where would that cropland been found? Would parks and woodland been cleared and cultivated? Would steeper slopes be cultivated and more wetlands drained? Could the United States export food, or would it be an importer of food?

"FREE" NITROGEN?

One method of obtaining nitrogen utilizes the growing of nitrogen-fixing legumes in rotation with grain crops such as corn. During one growing season the legume crop fixes nitrogen from the air, and that nitrogen fertilizes the grain crop as *green manure* during the next growing season. A legume green manure crop such as clover or alfalfa can capture between 80 and in some cases over 200 pounds of nitrogen per acre (90–224 kg N ha^{-1}) from the air. To secure that nitrogen as a green manure, the legume crop is *not harvested* but allowed to decay on site or most often plowed into the soil, and the nitrogen becomes part of the organic pool of elements in the soil.

Certainly this is a method of adding nitrogen to the soil for the next crop. However, although a green manure legume crop grown in rotation with food crops can add nitrogen, the other essential elements such as phosphorus and potassium available only from soil minerals are only recycled. No net gain results in the total quantity of phosphorus, potassium, or the other essential elements that originate in the soil minerals, although a transfer from the mineral pool to the organic pool may result in some increased availability of these elements.

A green manure crop rotation is not without cost to the farmer, who must pay property taxes and other fixed costs during a growing season in which the green manure crop is grown and no crop is sold. Also, additional land area needs to be cultivated to fulfill the demand for the food crop if a growing season is devoted to obtaining nitrogen fertilizer via a green manure legume crop rather than from chemical fertilizer.

Depending on individual circumstances, it may be desirable for a farmer to use a legume green manure option to obtain nitrogen. In recent years in the United States, the amount of nitrogen that could be obtained from a legume crop cost less when purchased as a chemical fertilizer than the taxes and other expenses involved in devoting a growing season to a legume green manure crop. Whether to buy the chemical nitrogen fertilizer or forgo a certain amount of cropland and include a legume green manure crop in the rotation is but one of the many management decisions that a farmer has to make.

Rotating food crops on individual fields is a common practice because it tends to interrupt the accumulation of crop-specific pathogens, and grain crops such as soybeans, wheat, and corn are often rotated for that reason. When speaking of crop rotation as a management option, a clear distinction must be made between a rotation that includes the growing of a legume that is not harvested (i.e., a legume rotation or green manure rotation) and simply planting different species of harvested crops in alternate growing seasons.

UNEVENNESS IN THE WORLD

The recent history of food production in the United States has not been attained globally. Greenland (1997) used Food and Agricultural Organization of the United Nations (FAO) data to determine that from 1950 to 1995 worldwide cereal crop yields have more than doubled, from 1.1 to 2.8 T ha^{-1} (approximately 16 Bu Ac^{-1}–42 Bu Ac^{-1}). Most of the yield increases have been in North America, Europe, and Asia while yields in some African countries have decreased over this same period of time (Figures 9–11 and 9–12). Decreasing grain yields have resulted in substantial amounts of native vegetation being cleared to make more land available for crop production (Figure 9–12). The effect of even slight yield increases appears to have slowed or stopped the clearing of land in some African countries (Figure 9–13).

In Africa approximately 70% of the people are engaged in farming. For many years farmers have been removing essential nutrients from their soil without replenishment with manure or fertilizer. Average annual removal rates are reported to average 19.5 Lb N Ac^{-1} (22 kg N ha^{-1}), 2.2 Lb P Ac^{-1} (2.5 kg P ha^{-1}) and 14.4 Lb K Ac^{-1} (15 kg K ha^{-1}). African farmers have adopted genetically improved varieties in about the same proportion as farmers in Asia, Latin America, and the Middle East. In those areas the improved varieties

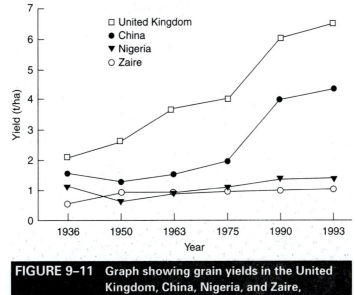

FIGURE 9–11   **Graph showing grain yields in the United Kingdom, China, Nigeria, and Zaire, 1936–1993 (T ha^{-1} × 893 = lb Ac^{-1}).**

Source: Greenland, 1997.

FIGURE 9–12   **Graph showing cropland area and grain yields in Niger and Sudan, 1970–1995 (kg ha^{-1} × 0.893 = lb Ac^{-1}).**

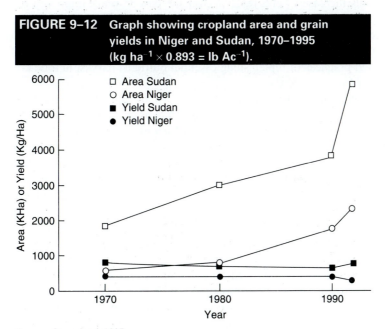

Source: Greenland, 1997.

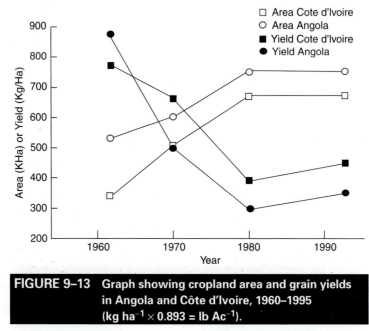

FIGURE 9–13   Graph showing cropland area and grain yields in Angola and Côte d'Ivoire, 1960–1995 (kg ha^{-1} × 0.893 = lb Ac^{-1}).

Source: Greenland, 1997.

account for a 60% to 88% increase in crop yields, whereas in Africa improved varieties have only increased yields by 25% over the past 38 years. Improved genetic capability is not achieved if essential nutrients are not available to the crop. Several social and political impediments in Africa have resulted in fertilizer costs that are two to six times higher than in Europe, Asia, and the United States, drastically limiting the ability of farmers to adopt accepted practices of replenishing essential nutrients harvest in food crops with chemical fertilizer (Sanchez, 2002).

Significant improvements to farmer yields have been made in parts of Africa by using locally available sources of phosphorus-bearing rocks and species of legume trees to capture nitrogen from the air. Much of Africa has only a limited growing season that corresponds to annual cycles of rainfall. One successful practice in parts of Africa is the planting of legume trees with corn. The trees are allowed to continue growing and accumulating nitrogen in their leaves for one or two years after the corn is harvested. After a couple of years the leaves can be harvested and applied as a residue to provide nitrogen and other essential nutrients they have accumulated when another crop of corn is planted in the same field. As the leaves decompose they provide available essential elements for the corn. The woody parts of the trees can be harvested and burned for cooking fuel. To provide phosphorus, certain phosphorus-bearing rocks are locally crushed with small grinding machines and applied to the soils. Although the phosphorus in the crushed phosphate rock material is less available to plants than chemically enriched superphosphate used in commercial fertilizers, it does provide some available phosphorus. When crushed rock phosphate is applied in concert with leaves of legume trees, the ambient meager corn yields are routinely doubled in parts of eastern Africa. By adding crushed rock phosphate and deliberately planting legume trees to grow during normal fallow periods, the frequency of food crop planting can be increased and the amount of land in customary slash-and-burn rotations can be reduced while obtaining the same amount of food production.

This methodology has several limitations. It has not been successful in western Africa where the dry season is too severe to support legume tree growth. In some areas a geologic source of suitable phosphorus-bearing rock is some distance from the culti-vated fields, and frequently even the relatively inexpensive rock-crushing equipment is not available. Additional labor is required to plant and harvest the leaves from the legume trees, and sources of legume tree seedlings are limited. Although this approach is considered low tech, it attempts to utilize available sources of nitrogen, phosphorus, and other essential elements without which benefits of genetically improved seed cannot be achieved (Sanchez, 2002).

"GOOD" SOILS KEEP GOING

Thus far this chapter has concentrated on soil use for cultivated food production. This use presently engages only about 11% of the total landmass of the earth. Estimates differ, but about 56% of the earth's land mass has little or no potential for crop produc-tion because of coldness, nonavailability of both rainfall and irrigation water, or lack of soil depth (Figure 9–14). Existing technology and economic conditions can seldom overcome these limitations. The remaining 33% of the land area includes about 10% of the land that is too wet to cultivate without engineered drainage and 23% with low quantities of chemical fertility (FAO, 1997).

FIGURE 9–14   Photo of nearly barren rock land in Arizona, USA.

A large portion of chemically infertile soil is present in tropical rain forests. As human demand for food increases, humans can either increase yields with fertilizer and modern management or invade the relatively infertile soils of the rain forests with low-yield slash-and-burn techniques and engineered drainage of wetlands. The course of events will depend on the course of local societal priorities. Individual farmers can neither determine nor control the many factors that determine agricultural production.

Societies' choices are much like those we make with our private automobiles. Like good cars that, with proper maintenance can be driven in parades of antique cars, good soils can keep supporting high yields of food production. There is the frequent need to fill up the soils' gas tank with the elements removed by the crops people harvest. Maintenance of tires, oil, and cooling systems must be attended to for a good car to keep running. Routine maintenance, like liming a soil to adjust for acidification or draining irrigated soils to remove salt buildup, is necessary when continuously using a soil for food crop production. Occasionally major repairs are required if a flood or severe drought occurs. But, like a good car, if constant care is given to the specific requirements, and these differ for each kind of soil, good soils will continue to serve as a media for any human use that is compatible with the characteristics of that particular soil. Even cars mangled in a severe accident and further crushed by a salvage company are recycled and provide material for new cars. Severely disturbed soil can be spread over the land to form a media for future plant growth as is done in the reclamation of surface mining sites. If we drive our soils more slowly, so to speak, with low food crop yields per acre, the need for nutrient elements is less than if we harvest high yields, but when only low yields are attained more cropland is needed to satisfy our need for human food.

Among the lands with potential for food crop production, some, like some cars, come with their gas tanks full and are ready to go. Most soils require some alteration before they are ready for crop production. "Dealer prep," it is called in the auto industry. Most land must be rid of natural vegetation much like a car dealer must wash a newly delivered car to rid it of the dust accumulated in transit and the inspection stickers left by the manufacturer. Some soils must be limed to increase the pH value to a level compatible with food crop production. Others require irrigation, drainage, or initial applications of calcium and phosphorus before a food crop can be grown.

All lands are not designed for all uses, any more than all cars are designed to be on the racetrack, in a drag race, or provide a luxury trip to the country club. Make and model of a car can often be correlated with the type and amount of maintenance needed. Similar comparisons can be made among different soils, and no universal formula can accurately predict when specific maintenance will be required. The judgment of the owner and operator who has daily contact with the machine must be relied on to take care of the situation. So it is with land and soil.

If we park a car no gas is needed, and it may appear preserved. Parks and wilderness areas can be established to preserve land. Even a parked car undergoes some change. A car may rust and require a new paint job even though it is never driven. A fire may denude preserved land, but a new canopy of vegetation will appear. To the human eye, parks and wilderness areas may seem preserved, but changes are taking place.

POSSIBLE SCENARIOS

With the scientific understanding of the chemistry of food crop production that has developed over the past hundred or so years, the future course of human food production does not necessarily have to follow historical patterns. In areas where there has been no improvement in crop yield per unit of land, the clear human response to the demands of increasing population is the clearing of more land for food production (Figure 9–12). In stark contrast, the development of high-yielding food production on the previously unproductive *cerrado* area in Brazil (see Chapter 8) is an example of what is possible when available scientific knowledge is utilized.

I have no crystal ball, but it appears that an entire range of possible scenarios will be played out in various parts of the world. Future agricultural practices will depend on the willingness and capability of individual societies to provide the essential components of agricultural production schematically represented in Figure 9–3. Failure to provide any one of the essential components of food production identified, and perhaps several others, will limit the total effectiveness and efficiency of land use for food production in various parts of the world.

Humans clearly have the power to alter natural ecosystems. Humans have a fundamental need for food. In the quest for human food, many natural ecosystems have been destroyed and the land used to better meet the requirements of humans. Whether future human requirements are met by intensive high-yielding food production on a small portion of the land available or by altering extensive land areas of natural ecosystems to grow low yields of human food crops will be determined by individual societies.

LITERATURE CITED

Andrews, W. B. 1954. The Response of Crops and Soils to Fertilizers and Manures (2nd ed., p. 463). College Station, MI: Author.

Anonymous. 1977. *Rothamsted Experimental Station Guide* (p. 79). Dorking, UK: Adlard & Son/Bartholomew Press.

Buol, S. W. 1995. Sustainability of Soil Use. *Annual Review of Ecology and Systematics* 26:25–44.

FAO, 1997. Accessed at http://www.fao.org/NEWS/FACTFILE/FF9713-E.HTM

Greenland, D. J. 1997. Inaugural Russell Memorial Lecture: Soil Conditions and Plant Growth. *Soil Use and Management* 13:169–177.

Grunes, D. L., and W. H. Allaway. 1985. Nutritional Quality of Plants in Relation to Fertilizer Use. In O. P. Engelstad (Ed.), *Fertilizer Technology and Use* (3rd ed., pp. 589–619). Madison, WI: Soil Science Society of America.

Larson, W. E., L. B. Nelson, and A. S. Hunter. 1952. The Effect of Phosphate Fertilization upon the Yield and Composition of Oats and Alfalfa Grown on Phosphate Deficient Iowa Soils. *Agronomy Journal* 44:357–361.

Lopes, A. S. 1996. Soils Under Cerrado: A Success Story in Soil Management. *Better Crops International* 10(2):9–15.

Mitchell, Professor. 1822, October. Portion of a speech delivered to the North Carolina Agricultural Society and published in the *North Carolina Department of Agriculture Monthly Bulletin,* No. 15. The author, identified only as Professor Mitchell, was most likely Professor Elisa Mitchell.

National Association of State Universities and Land-Grant Colleges. 1997. *American Agriculture: The Food, Fiber, Environmental System* National Association of State Universities and Land-Grant Colleges, Washington D.C. (p. 2).

Odell, R. T., W. M. Walker, L. V. Boone, and M. G. Oldham. 1982. *The Morrow Plots: A Century of Learning* (p. 22). Agricultural

Experiment Station Bulletin No. 775. Champaign: University of Illinois.

Pang, X. P., and J. Letey. 2000. Organic Farming: Challenge of Timing Nitrogen Availability to Crop Nitrogen Requirements. *Soil Science Society of America Journal* 64:247–253.

Rice Journal. 2004. Accessed at www.ricejournal.com/current/story

Russell, E. J. 1952. *Soil Conditions & Plant Growth* (8th ed., p. 635). London: Longmans, Green.

Ruthenberg, H. 1971. *Farming Systems in the Tropics.* Oxford, UK: Clarendon Press.

Sanchez, P. A. 2002. Soil Fertility and Hunger in Africa. *Science* 295:2019–2020.

USDA-National Agricultural Statistics Service. 2004. Crop Production Charts. Accessed at http://www.usda.gov/nass/aggraphs/cornyld.htm

van Straaten, P. 2002. *Rocks for Crops: Agrominerals of Sub-Saharan Africa* (p. 338). Nairobi, Kenya: ICRAF.

NOTE

1. The traditional numbers to identify chemical fertilizer are percentages of $N-P_2O_5-K_2O$. In this example, the fertilizer contains 17% of each. The P_2O_5 and K_2O are converted to elemental phosphorus and potassium by multiplying by 0.44 and 0.83, respectively. Thus each 100 pounds of 17–17–17 fertilizer contains 17 pounds of nitrogen, 7.5 pounds of phosphorus, and 14.1 pounds of potassium.

CHAPTER REVIEW PROBLEMS

1. As human population increases, does the land area used for producing food crops always increase?
2. In addition to good farming, what is necessary for efficient agriculture?
3. How long would it take you to carry an organic manure containing 0.2% phosphorus from a village waste site to your 1 acre of corn if that corn required 20 pounds of phosphorus per acre and was 1 mile from the village? (You have to carry the manure in a basket or other container along a walking path because you have no other way to transport the material.)
4. How long would it take you to transport the same amount of phosphorus as a super-phosphate fertilizer, 20% phosphorus content, under the same conditions outlined in question 3?
5. Why do trees and other native vegetation grow on land abandoned by farmers as too infertile for growing food crops?
6. What is the major consideration in the determination of the amount of fertilizer that must be applied annually to a field?
7. How do powerful tractors and other machinery contribute to the production potential of a skillful farmer?

10

Societal Obligations to Soil, Land, and Life

As discussed in Chapter 9, the efficient use of land for human food production has greatly improved over time in many, but not all parts of the world. In the United States (Figure 9–1) increased food crop yields per unit of land during the 20th century were clearly related to supplying essential nutrient elements via chemical fertilizers. Other innovations such as genetically improved cultivars and hybrids, better weed, insect, and disease control, mechanization for planting and harvest, increased use of irrigation on arid land, and improved drainage of land naturally too wet for maximum crop production have also contributed to increased yields. If scientific knowledge and technology exists, why have some parts of the world failed to increase yields of food per unit of land? The answer often rests with the economic, social, and political environment that surrounds the food chain.

SOCIETIES' INTERACTION WITH SOIL, LAND, AND LIFE

Discussions of human food production are often confined only to the role of the farmer. Many constraints limit a farmer's ability to utilize existing knowledge and technology. Irrigation can augment an inadequate water supply for crop growth and thereby greatly extend the food production capability of some arid and semiarid lands. Drainage can enhance crop production on land naturally too wet for good crop growth. Irrigation and drainage practices are seldom available to individual farmers without cooperation from the surrounding community. Individual farmers can seldom generate the power to operate the pumps necessary to move water. To replenish the nutrients taken from the soil by food crops, farmers may send grazing animals, such as goats and sheep, into adjacent lands to consume vegetation and return with the essential elements that vegetation contains in the form of manure to be used as fertilizer for food crops grown adjacent to their houses. This is not possible where surrounding land is not available for grazing.

Each growing season, a farmer invests labor and land equity in planting a crop that is expected to return some market value when harvested. If, for any reason, drought, flood, disease, insects, weeds, or no market demand for the crop at time of harvest, or the crop is not productive, the farmer loses part or all of those investments. Whether a farmer decides to increase his or her personal risk by investing in fertilizer, genetically improved seed, or any other equipment or technology that must be purchased depends on each individual farmer's perception of the amount of risk involved in that investment. A farmer is inclined to reduce risk by investing as little as possible in fertilizer or genetically improved seed without *reliable* and *timely* access to pesticides of the correct formulation to combat sporadic disease, insects, and weeds.

Perhaps the most pervasive risk each farmer takes is the reliability of a market for the crop that will assure the costs of production are recovered with a fair return for the labor and money invested. Without reasonable assurance that the costs incurred in the application of new technology will result in a benefit at the marketplace, a farmer is inclined to produce only what is needed to support the immediate family. *Reliable* and *timely* access to yield-improving technology and markets are societal responsibilities beyond the control of the individual farmer. Many parts of the world lack the political and economic stability necessary to develop the banking, transportation, and food-processing facilities necessary to alleviate the farmer's perception of risk and encourage use of yield-improving technology.

The entire food production system can be related to the links in a tow chain with which you are trying to pull a vehicle stuck in the sand. The amount of force that can be exerted through the chain depends on the power available for pulling and the strength of each individual link in the chain.

TECHNICAL-POLITICAL COMMUNICATION

With everyone in a society dependent on a reliable food supply, most governments attempt to play a role in ensuring food security. There is a fixed amount of land within each country, and all of that land is not of equal quality. Many, if not most, of the requirements for agricultural production diagrammed in Figure 9–3 are affected by government policies.

Governing bodies and other policy makers formulate incentives and/or penalties to influence farmers and other land users to adopt practices seen to be in society's best interest. Individuals in policy-making positions cannot be expected to understand the technical and scientific terms and concepts used by scientists to identify differences in land and soil. The spatial complexity of soil and land combined with the temporal uncertainties of weather challenge the simplification necessary for the formulation of land use policies. With a clear recognition that not all land will respond equally to any given technology or be equal in food production various land classification systems have been developed to identify the capabilities and limitations of land in terms that can be understood by policy makers. Land classification systems must be simple enough that individuals charged with applying policy directives can fairly identify and determine benefits expected to accrue in each category of land identified. No single land classification system is ideally suited for communicating all land uses to all policy makers, but one of the most used systems of land classification in the United States, and elsewhere, is the Land Capability Classification System (Klingebiel and Montgomery, 1961). The format of the Land Capability Classification System has been used for more than 40 years in the United States. There have been numerous technical changes in placing the more rigorously defined 22,000 kinds of soils recognized by soil scientists in the United States into the eight major categories of the Land Capability Classification System (Table 10–1).

TABLE 10–1 Land Capability Classification System[*]	
Class	*Limitations for Use*
I	Soils[†] with few limitations for use.
II	Soils with moderate limitations that reduce the choice of plants or require moderate conservation practices.
III	Soils with severe limitations that reduce choice of plants or require special conservation practices.
IV	Soils with very severe limitations that restrict the choice of plants and require very careful conservation management.
V	Soils with very little erosion hazard but are naturally too wet or subject to flooding to permit cropland use unless drained.
VI	Soils with severe limitations that make them unsuited for cultivated crops.
VII	Soils with very severe limitations that make them unsuited for crops or managed pasture.
VIII	Soils and nonsoil landforms suited only for wildlife and recreational uses.

[*]Numerous technical documents specifically define soil and landscape features that allow land use technicians to place each of the 22,000 different soils recognized in the United States into one of these classes.

[†]Soil as used in this system also includes landscape features such as slope.

Four subcategorizes in the system identify specific conditions that either limit use of the land or require specific land management techniques for each major category, except Class I land that is defined as having few land use limitations. Land that is susceptible to erosion because of slope is designated by *e* (i.e., II*e*, III*e*, etc.). Land where some uses are limited by wetness or flooding is identified with *w* (i.e., II*w*, III*w*, etc.). Land where the rooting depth in the soil is limited by the presence of rock, low moisture-holding capacity, or other features are identified by *s* (i.e., IV*s*, etc.), and land where food crop production is limited by climatic conditions of temperature or moisture are designated with *c* (i.e., VII*c*, etc.).

In the Land Capability Classification System, class I land is nearly level with deep soils. None of the conditions identified by the subcategories just described are present. With proper attention to maintaining fertility and soil pH, Class I land can be continuously used for intensive production of food crops. On Class I land the greatest productivity can be obtained with the lowest costs of production.

Class II land can also be used for intensive food crop production, but some practices such as strip crops and grassed waterways should be used on erosion-prone slopes, subcategory *e* land (Figure 10–1), and some engineered drainage is beneficial in subcategory *w* land.

Class III land is more susceptible to erosion or wet conditions than Class II land and requires more narrow spacing of strip crops or drainage systems in subcategory *e* and *w*

FIGURE 10–1 Photo of land capability class II*e* land in Nebraska, USA.

land, respectively. Although satisfactory yields can be obtained with proper management, the cost of production on class III land is higher than on class II and I land.

Class IV land can be used for food crop production where intense erosion control practices such as terraces on *e* subcategory land or intense drainage is installed and maintained on *w* subcategory land. Production costs are higher because of the need to install and maintain either terraces or drainage.

Class V land is generally not suited to food crop production because of shallow rooting depth, an extremely short growing season, or frequent flooding. Grazing or timber production is possible, although such uses may be limited by one or more conditions (Figure 10–2).

Class VI land is not considered useful for crop production but satisfactory for timber harvest, grazing, or wildlife.

Class VII land has severe limitations for grazing or timber production because of steep slopes, shallow rooting depth, or short growing seasons. Timber production is low and extra care is required in harvesting timber on steep subcategory *e* land to reduce erosion, and grazing is limited by the slow growth of grasses (Figure 10–3).

Class VIII land should not be used for any kind of commercial production. It has value for wildlife, recreation, and aesthetic uses. Much of the class VIII land has very steep slopes. Where possible, maintaining good vegetative cover reduces erosion and protects the quality of runoff water that fills reservoirs and lakes (Figure 10–4).

The intent of Land Capability Classification is to identify the site-specific characteristics of land and convey those characteristics in a format that can be understood by policy makers and those charged with administrating land use policy. Better communication

FIGURE 10–2 **Photo of tidal marsh, land capability class V land in North Carolina, USA.**

between scientists and policy makers helps ensure that expenditure of public money is applied to the kind of land that will benefit from that expenditure. It is unrealistic to expect political entities to have the knowledge of land conditions or the ability to establish programs that related to land use except through a system that identifies broad groups of land conditions that could benefit most from the intent of a specific program. There is no single land use practice that can be encouraged by incentives or discouraged by regulations that can equally benefit all land. Society is ill served if expenditures are made for programs that attempt to alter land use on land not likely to benefit from a particular land use incentive program or regulation. Throughout the United States, soil scientists with knowledge of local land and soil conditions have classified each of the more than 22,000 different kinds of soil into one of the land capability classes.

In some developing countries, governments have used the Land Capability Classification System and several other similar land classification systems to determine where roads should be built to access the greatest amount of class I and II land and enhance food production in their country and discourage building roads into potentially less productive areas. Land classification systems also provide agribusiness with information as they seek to locate marketing centers and retail outlets. Agribusiness suppliers of specific agricultural supplies such as fertilizer, pesticides, or machinery locate retail facilities in areas where there is adequate demand for their products. Marketing and processing facilities are located in areas where the land is capable of producing the products they buy from and sell to the farmers.

FIGURE 10–3 Photo of land capability class VII land in Arizona, USA.

The concerns and priorities of societies change with time. These changes are often reflected in land use policies. In the United States following World War II, the need for food production both for domestic use and export was great. During that time farmers could receive financial aid from the federal government to install drainage ditches and tile lines in classes IIw, IIIw, and IVw (i.e., wetlands), thereby increasing the amount of land that could contribute to food production. During the 1970s, abundant food was being produced in the United States, and societal concerns became focused on environmental protection (Helms et al., 1996). In response, most subclass w land was designated as wetlands, and federal regulations were enacted preventing landowners from installing additional drainage systems, although existing cropland with established drainage systems was allowed to remain in production.

The Conservation Reserve Program (CRP) was enacted in 1985 and aimed at retiring land most susceptible to erosion from crop production through ten-year contracts between the federal government and landowners. Land identified as Land Capability subclass e could be taken out of food crop production and rented to the federal government as conservation reserve land. In many parts of the United States, there is concern that urban development is excessively encroaching onto some of the best farmland. To curtail this encroachment, legislation was enacted that declared class I and some class II land as "prime farmland." No financial loans could be obtained from government sources for urban development on such land.

FIGURE 10–4 Photo of land capability class VIII land on the banks of the Yellowstone River in Yellowstone National Park, Wyoming, USA.

POLICIES TO PROMOTE SOIL CONSERVATION

Erosion control has broad societal benefits. In the United States federal funds have been used to help farmers install proper erosion control practices because clean water benefits all of society. In 1933 the federal government created the Soil Erosion Service (SES) to combat soil erosion in the United States. Amid jokes that the agency "serviced soil erosion," the name was changed to the Soil Conservation Service (SCS) in 1935 and in 1994 renamed the Natural Resources Conservation Service (NRCS), within the U.S. Department of Agriculture (USDA). In cooperation with land grant universities, the USDA established research sites throughout the country to study methods of farming that reduce soil erosion and demonstration projects to teach farmers soil conservation practices.

Research efforts were instrumental in developing a systematic method of estimating rates of soil loss known as the universal soil loss equation (USLE) (Wischmeier and Smith, 1978). The USLE was used to estimate rates of sheet and rill erosion expected to occur when specific farming practices were used on various types of soil and sloping land. The USLE served as a guide for recommending alternative cropping systems, management techniques, and conservation practices on erosion-prone land identified in the Land Capability Classification System.

One of the areas where erosion was of greatest concern was the upper Mississippi River watershed in parts of Minnesota, Wisconsin, and Iowa. That area has thick naturally fertile soils formed in loess. It is also an area of steep slopes, and in the 1930s, excessive erosion was clearly evident in farmers' fields. A study has been made to determine the changes that have taken place and evaluate the effectiveness of soil conservation efforts in five representative counties in that area from 1930 to 1992 (Argabright et al., 1996). The study found that average erosion rates from cropland in the five counties decreased from 14.9 tons $Ac^{-1} Yr^{-1}$ (33.4 Mg $ha^{-1} Yr^{-1}$), in 1930 to 7.8 tons $Ac^{-1} Yr^{-1}$ (17.5 Mg $ha^{-1} Yr^{-1}$) by 1982 and 6.3 tons $Ac^{-1} Yr^{-1}$ (14.1 Mg $ha^{-1} Yr^{-1}$) in 1992. This reduction in erosion took place although the land area used for row crops, small grains, or rotation meadow was 16% greater in 1992 than in 1930. The reduction in cropland erosion was attributed to investments in terraces, contour strips, and reduced tillage. The study also noted that in 1994, 18% of the most erosion prone cropland in the region had been enrolled in the CRP.

The authors (Argabright et al., 1996) suggested that several changes in farm management practices that had taken place from the 1930s to the 1990s had contributed to reducing erosion in addition to physical erosion control practices on the land. For example, in 1930 tenant farmers harvested 34% of the cropland, whereas in 1992 tenant farmers harvested only 12% of the cropland. Horse, mule, and sheep numbers sharply declined since the 1930s, decreasing the amount of grazing in woodlands. As stated by the authors, their study represents only one area, and the results cannot be confidently extrapolated. The potential to extrapolate is limited because of the paucity of quantitative of long-term data on erosion. The study clearly demonstrates that effective erosion control practices are known and acceptable to farmers who as landowners make the final decisions about the management of their land in the United States.

IMPEDIMENTS TO FOOD PRODUCTION

Several of the components required for high levels of agricultural production (Figure 9–3) have failed to materialize in many parts of the world. No single scenario can be identified to account for the failures. An almost infinite number of reasons are possible, each somewhat localized to specific countries or regions of the world. The following experiences are but a few examples of scenarios that impede agricultural production in many parts of world.

In one country the government officials were being approached relative to a U.S.-funded research project on soils. The objective of the project was to study a land area that encompassed nearly half of that country. At that time there were very few inhabitants in the area, reliable roads and other transportation systems to the area did not exist, and the area produced almost no food crops beyond the subsistent needs of a few people living there. A high-ranking government official opposed the project. The U.S. scientist, with the support of some local scientists, pointed out that the region had adequate temperature and rainfall to sustain production of food crops. The area was nearly level and not subject to flooding or erosion. The reason for the lack of production was postulated to be acid soils with low phosphate content. Data were presented demonstrating that similar soil conditions in other countries had been successfully corrected with proper soil management and were now producing substantial amounts of food. There was a high probability that if the project was successful, this large area of the country could be a significant food source for his country that at the time imported large quantities of food. The reason given for his objection was that there were few people in the region. The scientists acknowledged that fact but argued that the project was not to improve the productivity of indigenous inhabitants but offered long-term development and productivity for the whole country. Still objecting to the project but assuring the scientists that a U.S.-sponsored project was welcome, the official suggested that the project be located in the high mountainous region of the country where population density was great and crop yield low. The U.S. scientist assured the official that he was well aware of the situation in the high mountains and was convinced the low yields were due to year-long cold conditions in the high elevations, and soil scientists had no potential solution that could correct low temperature. When the official was pressed for a further explanation of his objection, he offered the following scenario. We are a democratically elected government. If we have a visible U.S.-sponsored agricultural project in those parts of the country where there are a lot of votes, we (his particular political party) enhance the chances that we will again be elected. If the U.S. project was located in an area where there are few voters, there would be little political benefit. We, as a political party, need to show attention to the immediate benefit of our people if we are to be reelected. It is difficult for a scientist to argue the pragmatic realities of political life.

In another country, a particularly disturbing scenario was encountered. Local agricultural scientists, many with graduate degrees from universities in Europe and the United States, insisted that the techniques of farming they were taught and observed as students would not work in their country. Evidence for their contention was readily seen throughout the country where large fields had only sparse populations of grain crops and erosion was evident in many fields where the land was even gently sloping. These grain fields had been in production for many years but were now failing. It was obvious that something was wrong. When there was an opportunity to talk in private and away from the ears of government officials that may report him as antigovernment, the owner and operator of one farm where the crops were failing offered the following scenario for the failure of his crop. He related that he and his father before him had farmed there for many years. In recent years the government had forbidden expenditure of hard money outside of the country. He went on to relate that he had American dollars, Italian lire, British pounds, and French francs, but when the expenditure of hard money outside of the country was forbidden, he could no longer import tires for his tractors and combines, and tires were not made in that country. He went on to say that at planting time fertilizer was no longer available, and he could not import hybrid seed so had to plant locally available seed that did not have good genetic potential. He knew from past experience that modern farming practices were successful but impossible under the present political conditions. The most disturbing aspect of this situation was that the local agricultural scientists had failed to diagnose the cause of the crop failure. They insisted that new scientific methods had to be developed to grow crops in the area because agricultural science from the developed world did not work in their country.

In another less repressive country, the local village council of a small village where the people had been farming the same fields for generations was assembled to answer questions about their crop production. When asked what yields they were obtaining in their best fields, they responded "about 200 kg ha^{-1}" (about 200 lb Ac^{-1}). Such a yield is so low, it barely recovers the amount of seed planted. When asked what the highest yields they had ever gotten on their fields, they responded by looking into some handwritten records and came up with a figure of about 2,000 kg ha^{-1} (about 2,000 lb Ac^{-1}) and indicated that those yields were some 25 years previous. When asked why they thought the yields had declined to 10% of their former level, they chuckled and responded, "At that time we were a colony, and chemical fertilizer was imported. Fertilizer is no longer available."

The easy and common knee-jerk reaction to low food production is to look for problems on the farm and often fix blame on the ignorance or reluctance of the farmers to adopt new practices. It is very disturbing to find a chasm of understanding that sometimes exists between agricultural scientists and farmers. In one fairly well-to-do country, farmers were observed threshing grain with a tractor-powered threshing machine that only beat the grain from the straw and then deposited both the grain and straw onto a canvas. From there the farmers, and their entire families, separated the straw with a hand fork, placed the grain and chaff into large shallow baskets, and tossed it into the air so the wind could winnow the chaff from the grain. Power-driven threshing machines that separated the grain from the straw, winnowed the chaff, and delivered clean grain were precursors to combines and widely used in Europe and the United States in the early 20th century. The machine they were using was only half of the now obsolete threshing machines. When the local agricultural scientists were asked why only part of a threshing machine was being used, and it was evident that tractors and other rather modern machinery was available, they responded, "The people enjoy winnowing the grain by hand." It was over 100°F (38°C) at the time, and one could seriously doubt that winnowing grain was an enjoyable task.

It is common to find that agricultural scientists in developing countries have little or no personal experience with farming. A probable reason for this is the lack of elementary education in rural areas, thus limiting the chances that children with practical farm experience can ever advance in the educational system and become agricultural scientists.

A seldom mentioned but perhaps all too common reason for political leaders to ignore the development of infrastructure required for productive agriculture was succinctly stated by an official of a particularly repressive government as follows. When our people are scattered throughout the land, each producing only the amount of food their families require, they cause the government few problems. When the people congregate in cities, they form groups that riot and create problems. Clearly, some governments discourage productive agriculture that could free many of their citizens from subsistence farming.

Benevolence on the part of well-meaning countries and groups to help ill-fed people is sometimes cited as a cause of neglecting agricultural development in some countries. Government officials in countries receiving food from benevolent organizations often point out that it would be unfair to expect their urban people to pay more to purchase food from local farmers when food is made available from outside sources at a cost below that of local production. Therefore they see no need to invest in agricultural research and development that would benefit their farmers.

AGRIBUSINESS AND AGRICULTURAL PRODUCTION

In all but the most primitive forms of subsistence farming, all businesses deal with agriculture. For example, the steel industry is involved in the making of a machete to be used by slash-and-burn farmers. The term *agribusiness* is often confined to those enterprises that deal directly with buying, selling, and trading of agricultural-related commodities. Agribusiness is present at all size scales from the direct marketing of food at an open market (Figure 10–5) to international companies that import and export food and farming supplies throughout the world (Figure 10–6).

FIGURE 10–5 Photo of an open market in Cuzco, Peru, SA.

FIGURE 10–6 Photo of a Pan Am Cargo 747 being loaded in São Paulo, Brazil, SA.

The most basic requirement that farmers have before they attempt to produce food in excess of the needs of their family is a reliable market. The growing of every food crop requires time. Most grain crops are harvested about 100 days after planting. A much longer period of time is needed for investments in meat and milk animals to be productive. Even longer periods of time are needed for trees to produce fruit and nuts or valuable wood products. Few farmers are able to individually finance required inputs such as genetically improved seed and fertilizer for grain crops, the raising of mature dairy or meat animals, or the planting of fruit trees. Agribusiness includes the credit institutions that risk the capital necessary for farmers to initiate the production of food for sale.

Farmers attempting to convert from subsistent farming to production farming also incur a personal risk that if they convert their land to the production of commercial crops and divert their labor and land from growing a variety of food crops to feed their immediate family, they may find their families without suitable food. Thus a critical component of agribusiness is providing not only a market for the farmers to sell products but also a reliable source from which farmers can obtain food previously grown on their land.

All business ventures involve risk. The marketing side of agribusiness has three especially risky components because of the nature of buying food products from farmers:

1. Many food products are perishable and time is critical. Also, many food grains are produced seasonally, but the market demand for food is not seasonal. To accommodate the seasonal nature of supply, storage facilities and processing plants to preserve the food products are vital. Capital investments must be made in storage facilities that are only seasonally used.

2. The source of farm-produced food products is extensive often over large distances. Reliable rapid transportation is required to minimize spoilage (Figure 10–7). In most countries governments are responsible for building and maintaining transportation systems. Agribusinesses avoid countries that fail to support reliable transportation systems.

3. The supply of agricultural commodities such as grain, meat, or fruit is subject to interruptions because of weather or disease. The flow of food products is not as steady and smooth as the flow of automobiles on weather-protected assembly lines.

The supply side of agribusiness is faced with the same distribution and seasonal nature of farmer demands for their genetically improved seeds, pesticides, and fertilizers. Lime, that indispensable ingredient needed to maintain pH values in acid soils so that it is maximum availability of essential elements is obtained, is a heavy product. Limestone rock is frequently distant from areas of naturally acid soils. The crushing of limestone into fine particles so that effective in raising the soil pH does not require especially large or expensive equipment. However, the transportation of adequate amounts of lime, from 1 to 5 or more tons per acre (2,240–11,200 kg ha^{-1}) of cropland, presents difficult problems without adequate rural roads.

The mining of phosphate- and potassium-bearing rock and the manufacture of concentrated nitrogen, phosphorus, and potassium fertilizers requires large capital investments, expensive equipment, and a reliable source of chemicals and power. Sources of phosphate- and potassium-bearing rock are present within every country in Africa (van Straaten, 2002). The need for concentrated fertilizer in Africa is

FIGURE 10–7   **Photo of a raft loaded with farm products docking at Yurimaguas, Peru, SA. The farmer and his family will attempt to sell everything, including the logs, before they make their way back to their home in the remote jungle.**

great, but the lack of stable trade agreements that would assure the companies manufacturing fertilizer an adequate market beyond the borders of the country where the facility was located has deterred construction of fertilizer manufacturing facilities. In some countries, geologic deposits suitable as a raw material for manufacturing concentrated fertilizer are limited and do not presently justify the expense of constructing fertilizer-manufacturing facilities. The total known geologic deposits containing concentrations of essential elements suitable for refining into fertilizer are finite. However, from known proven reserves and estimates of potential use, the possibility of total worldwide exhaustion appears remote.

THE RIGHT PLACE AND THE RIGHT TIME

The multitude of different soils present in the world precludes any one technology or policy from providing a single solution for sustained food production. Although the general technology of fertilization has been responsible for dramatic increases of food production per unit of land on certain soils, it is not applicable on many others. Fertilizer will neither warm a soil that is too cold nor irrigate a soil that is too dry. It does no good to supply large quantities of fertilizer if the genetic capability of the crop plant limits potential production. It does little good to fertilize a soil that is so acid that crop plants find it impossible to extend their roots. *Fertilization* and *fertilizer* are vastly oversimplified terms. To be effective, fertilizers must have specific formulations that meet the needs of the specific crops and be applied at the specific site and time that the crop is grown. Fertilization, the act of applying fertilizer, must include amendments that adjust soil pH to a range compatible with the crops grown. Efficient fertilization must include placement of the fertilizer and other needed amendments at the proper depth in the soil. Physical manipulation of the soil may be required so that hardness does not preclude root elongation. Fertilization, inclusive of all the cultural practices involved in good soil husbandry, must be carefully orchestrated for each site and for each cropping season. Some fertilization practices will be effective for several crops, whereas others must be repeated and/or modified for each crop. The eye of the farmer, as extended to include the sensory capability of chemical soil analysis, is critical in effective fertilization.

It has been said that the weakest part of an automobile is the "nut" behind the steering wheel. It is also said that the same nut is also the most dangerous part of the automobile. So it is with the utilization of land and soil, and the farmer is that nut. When food is not plentiful, cheap, and readily available, we tend to blame the farmer. But just as most automobile drivers are good, most farmers are good at what they do. Like most automobile drivers who adjust their driving practices to conform to laws and traffic and road conditions, most farmers drive their agricultural production according to their perception of economic and physical risk. Indigenous farmers have learned how to drive the soils they have available within the technology, social, economic, and political structure that surrounds them. A good driver would not attempt to use rush hour driving skills while on a scenic roadway. Should farmers be expected to produce food for the whole of society if land is socially appreciated only for scenic aesthetics? We would not expect a model-T Ford to compete in the Indianapolis 500. Should we expect farmers to attain high yields in a society where genetically capable seed, fertilizer, and machinery are unavailable? We would not expect a person with experience only with a small economy vehicle to drive a souped-up stock car in the Daytona 500. Should we expect farmers to acquire new skills where rural education is so limited that few learn to read and write? Effective use of technology to improve crop yields requires not only the availability of the technology, but the driver (farmer) must acquire the skills in the utilization of that technology and receive societal support that assures the well-being of farm families.

SUSTAINABILITY

Attempting an analogy between automobiles and soils is perhaps unfortunate in that we tend to think of an automobile as something that will wear out and be replaced. The existence of vintage automobiles attests to the fact that they do not wear out if there is complete maintenance and repair. So it is with soils. Not all soils need the same maintenance, but there are several commonalities. Replenishment of chemical elements extracted in crops and sold as food in distant cities can be equated with filling the gas tank of an automobile. Without assurance of a reliable gas station along the way, we cannot drive very far. If the quality of the fuel we buy is not compatible with the needs of our car, it will cease to function. Without reliable replenishment of essential elements, we cannot indefinitely harvest food crops from any soil. Just as more sophisticated and powerful cars require more specific fuel and maintenance requirements, obtaining higher crop yield requires more specific inputs and careful management.

Soils can be destroyed by erosion to bedrock or buried in a volcanic eruption just as an automobile can be destroyed in a fire or by a severe crash. But just as the twisted metal remaining after an automobile has been wrecked can be melted and again machined into parts for a new automobile, land denuded by erosion still has an interface between the geologic mineral substrate, the atmosphere, and radiation from the sun. Soil will again be present—not the same soil as before but a soil.

An automobile can be wrecked in several ways. When two or more cars collide, it is usually the fault of only one driver. An automobile can be crushed as a tree falls in a tornado. That is an act of nature, and no one is at fault. In like fashion, some farmers destroy some land, but not all farmers destroy land. Acts of nature have been destroying and creating soil long before humans arrived on the scene.

Nature's time scale for destroying and rebuilding soil most often exceeds human concepts of time. For an example of the natural destruction and formation of land, we can observe Chimney Rock, that milepost along the Oregon trail that signaled to the pioneers traveling in wagon trains that at long last they crossed the endless prairie on the way to Oregon (Figure 10–8). Note that the chimney-like spire is not a rock but the last remnant of an old land surface. Its sides are barren and exposed to further erosion that has provided material for a new land surface that surrounds the chimney. The bales of hay in the foreground of the photo clearly demonstrate that the new land surface is playing a part in the production of food.

FIGURE 10–8   Photo of Chimney Rock, Nebraska, USA.

A STITCH IN TIME: PREVENTATIVE MEDICINE

Although catastrophic events of nature such as continental glaciations and volcanic eruptions are in part responsible for the formation of many kinds of soils, the human time scale is not well attuned to the geologic time scale. Human perception of risk determines human actions. The risk of being without food and water for even a very short period of time is fundamental and universal. Only when the risk of hunger and thirst is reduced by the development of agricultural systems that reliably deliver food and drink can human energies be directed to other societal concerns.

Concern for conservation of the environment is probably the first casualty in the mind of a farmer with a family facing hunger. In the semiarid lands of the Sahel in West Africa south of the Sahara Desert, cattle herding has been a way of life for untold generations. Climatic and vegetative conditions are much like those in west Texas in the United States. Like west Texas, the amount of rainfall is the most critical variable that a rancher in the Sahel faces each year. In Texas, good sustainable range management is achieved when the number of cattle allowed in an area of rangeland is kept low enough that some of the most desirable grass species are not eaten even during the driest of years. This assures that desirable grass species will survive and provide good grazing in years to come.

An experienced range conservation scientist from west Texas went to the Sahel to advise on grazing practices and proceeded to teach this practice to the tribal herdsmen in the Sahel. He had very limited success. The herdsmen in the Sahel attempted to have a large enough herd that the cattle would consume all the annual grass in the years of abundant rainfall. The rationale for their management was explained to the range conservationist as follows. If I (the Sahel herdsman) leave grass on the land during a rainy year and next year there is a drought, my family and I cannot eat that grass. If I "put hide and hair around it" [allow the cattle to eat all of the grass in a rainy year], my family and I will have something to eat should a drought occur next year. Ranchers in Texas most likely can rely on a stable financial system for a loan to see them through a bad year and buy cattle to replace those sold to conserve desirable grasses during dry years. Perhaps they can find some part-time employment for a little extra cash. The herdsman and his family in the Sahel were totally dependent on providing themselves with food. The range conservation scientist found it difficult to disagree with the Sahel herdsman once the reasons for his practices were explained. Social and economic conditions clearly affect a farmer's or rancher's land use decisions.

Secure and reliable food supplies are obtained via a multitude of pathways within the world today. The more diverse the available pathways, the more risk free the supply of food. Within a society with a multitude of avenues available for food production and delivery, the effect of sporadic events such as flood and drought are mollified. Reliable transportation of food has become a key element in the utilization of diverse soils for reliable food production. Reliable storage and transport of food has allowed societies to concentrate the production of certain food crops on soils that are most compatible with the needs of specific crops and develop the economic advantages of large-scale production. Storage, processing, and centralized marketing systems reduce waste that occurs when producers sell only to local consumers. All of these components of the food production chain help reduce risk.

In areas where reliable markets for specific crops are established, farmers specialize and become highly skilled in management of specific crops. Investments in fertilizer, seed, and pesticides are relatively risk free if they help ensure a high yield and a reliable market is available. The investment in specialized machinery to plant and harvest only a few varieties of crop plants is less than when farmers grow many different crops on their land. Only when farmers have confidence in long-term security will they make the substantial investments involved in specialized crop production and have concern for the conservation of the land resource.

Where reliable markets are not available, farmers limit their objectives to reducing risk and attempt only to feed themselves and their immediate family by planting many types of crops with low inputs of fertilizer and little attention to genetically improved seed. High yields are less important than reliable yields. Subsistent farmers become skilled in selecting specific sites to plant specific crops and effectively use the diversity of soils within small geographic areas. This locally acquired indigenous knowledge is attuned to reducing food production risks. In the absence of reliable storage facilities, selection of crops grown is also dictated by the necessity of a continuous supply of food. Long-term concerns for soil conservation are of low priority for subsistence farmers when the family food supply for next week or month is uncertain.

The choice between specialized crop production and subsistence farming is not available in all societies. Subsistence farming is the only alternative available to people in those areas of the world where financial and/or political instability precludes investments in reliable rural transportation, food storage and marketing facilities, and other business infrastructure necessary for specialized farming. In some countries more than 80% of the people are relatively immobile subsistence farmers who produce little or no food for sale but seek only to feed their families. These people have to obtain almost all of the food in their diet from a limited range of soil properties within the local area. Most often technologies that could replace existing techniques are limited by social, economic, and political constraints beyond the control of individual farmers. The underlying causes of these constraints are too numerous to attempt more than a cursory examination. In some cases political rulers feel threatened by the concentration of people in cities and prefer to keep as many people as possible dispersed and less likely to organize and cause trouble. Some locations that have land well adapted to intense food production are isolated by difficult terrain that hampers the construction of reliable roads, and some are simply too distant from population centers to make marketing of food products profitable. There is often little or no incentive for government leaders to attempt improvements in transportation to these areas simply because there are no votes to be gained by spending money that serves only a few, often politically inactive, people in remote areas of a country. In some cases international supplies of food can meet metropolitan consumer demands for food at a lower cost than farmers within the country.

SOCIETAL EXPERIENCE WITH FARMING

Some systems of government have attempted to improve the quantity and reliability of human food production through centralized management of farming operations. The concept is appealing. In concept it is reasoned that if the best available data relative to the best planting time, the best fertilizer rates, the best cultivation practices, the best pest and weed controls, and so on, were averaged for a region and applied uniformly, improved yields would result. The results of this concept became evident after several years in the communist-controlled areas of Eastern Europe and Asia. Centralized control of farming ignores two key ingredients of farming. First, not all the land within an area has the same kind of soil. Uniform management for the *average* soil is not the best management for *all* soils. Second, the assembly line for food production is not protected from the weather. Attempting to enforce compliance with planting dates and weed and pest control applications resulted in local production problems every place and every year when *average* weather conditions did not materialize. For example, in the foreground of Figure 10–9, wide spaces can be seen between the soybean plants within the large field of land capability class I and II land. The director of this large commune farm within the Eastern European bloc of the former Soviet Union lamented that the poor establishment of the soybean crop was due to the directive he received ordering that soybeans had to be planted by a specific date. In that particular year, exceptionally large amounts of rain had fallen for several weeks before the prescribed planting date. Fearing reprimands if he failed to comply with the directive, he had ordered that plowing and planting take place even though the soil was too wet. The wet soil stuck to the equipment and the distribution of seed was erratic, resulting in an irregular and sparse spacing of the soybean seed. Although in the distance the field appears very healthy, there can be no doubt that potential yields were not obtained with the low plant population that resulted from planting when the soil was too wet.

Many governments have attempted to disperse food production from large privately owned farms by land reform programs that placed a large number of persons, inexperienced or experienced only in subsistence farming, into ownership and operation of land that was expected to produce commercial food crops. Almost without

FIGURE 10–9   Photo of a soybean field on a state farm in Bulgaria. Note the sparse stand of soybeans caused by planting when the soil was too wet.

exception these perhaps socially desirable experiments have proven to be failures. The inexperienced land reform farmer lacked the business expertise to manage credit from planting to harvest and/or the banking facilities lost confidence and failed to serve the needs of the small farmer. Also, the small size of each farm negated the efficiency of mechanization available to larger farms. The science and art of timely fertilizer application, irrigation, cultivar selection, and pest and weed control was often badly conducted by the inexperienced farmer. Marketing became more difficult as agribusinesses were forced to deal with several individuals within the same production area.

THE ART OF SOIL, LAND, AND LIFE

Just as there is no single food favored by all people or single type of music that will please all audiences, no single system of management is best for all soils. Each type of soil has a range of management systems and practices within which it will continue to function as a media for the growth of human food plants. Subtle differences in soil are present within every area of land. Much like the audience at a concert, some soils will become excited and respond with loud applause (high crop yield) to a Sousa march but only politely applaud (a low crop yield) to a Beethoven symphony. Astute musical directors will select specific music for a concert based on experience and anticipated audience.

When allowed freedom to make individual decisions, astute farmers will select management practices compatible with individual soils they cultivate *and* the reliability of economic, social, and political infrastructure that surrounds them. Just as the mood of an audience may change during the course of a performance, soils change with each day, becoming wetter with each rain and drier during rainless days. The physiological stage of a crop daily changes from seedling to maturity. Market prices and transportation costs often change almost as rapidly. For efficient and safe food production, an astute farmer will respond to all of these changes.

Preceding chapters were devoted to the physical and chemical necessities of soil and land and their interactions with food production necessary for all forms of life. Much attention was given to the questions of *what* farmers had to do to assure production of human food crops. The determination of *what* is needed to assure food crop production has been the focus of numerous research projects by agricultural scientists, both public and private, throughout the world. Numerous research reports, extension guides, and books covering topics of insect and weed control, fertilization recommendations, cultivation techniques, animal care, marketing strategies, and so on, have been written.

An equally important consideration for efficient food crop production is concern for *when* management operations should be conducted. In scientific research the *when* can only be determined within the confines of the conditions that prevailed during the experiments. Recommendations to farmers can be presented in terms that describe the conditions of *when* a management practice should be conducted are often based on average or normal weather conditions within an area. The correct although amorphous definition of *when* a specific management practice should be conducted is "on time."

The correct time to conduct farming operations differs spatially depending on the type of land and soil and hourly on weather conditions. Here is one apparently subtle question that reveals an individual's knowledge and experience with harvesting soybeans: "What time in the morning do you begin harvesting soybeans with a combine?" An answer such as 8 o'clock or 10 o'clock reveals unfamiliarity with combining soybeans or any other grain crop. An experienced farmer will answer, "When the dew is off." This answer is based on the knowledge that the grain will not be cleanly separated from the stems and chaff if it is too wet. After moisture has condensed on the plants as dew during the cold of the night, soybean and other grain plants are often too wet for efficient operation of a combine. How many minutes or hours it will take for the dew to evaporate depends on the amount of dew, the relative humidity, the wind speed, and the temperature that morning. A correct time of *when* to start the combine can only be determined by directly observing the field each morning during the harvest.

Although greatly aided by chemical measurements that extend the observational abilities of the farmer, the art of good land use, whether for farming, ranching, lumber, or recreation, always requires that the correct management practice be applied at the right place and at the right time. In this regard Aristotle is reported to have said, "The mark of a good farm is the footprint of the owner in each field each day." The author believes the point Aristotle was making is illustrated in the following scenario.

Several years ago an experienced manager of a large corporate farm in the southern part of the United States was asked, "Will large corporations take over farming in the United States?" He emphatically answered "no." When pressed for the reasons for his negative answer, he replied with this scenario. A corporation hires a bright young agricultural graduate to manage 1,000 acres (405 ha) of corn and soybeans. After the first crop season there are excellent crops and the company has a big profit margin. At Christmas the young manager is given a big bonus, a substantial increase in salary, and told, "You are exactly the kind of person this company is looking for and you will advance within the company." Next year you will have 2,000 acres (810 ha) to manage." The next year the young manager works very hard and again big profits are realized. At the next Christmas there was again praise, a big bonus and a big increase in salary. The assignment for the next year was, "You will have 5,000 acres (2,025 ha) to manage." The young manager is capable and efficient, but the on-site observations necessary to make timely decisions are not possible on such a large area. At the end of the crop season there are big economic losses. In view of that performance, the company determines that the manager will have to be fired.

Quite simply, one decision maker, in this case the young manager, cannot physically evaluate the site-specific conditions necessary to make timely management decisions on such a large area. Timing of management operations is critical for efficient food production. Largely unpredictable changes in the weather, superimposed on differences in individual kinds of soil, outbreaks of disease and/or insects, and the growth characteristics of crops require continuous, site-specific alteration of management operations. Failure to protect strawberries from freezing conditions that may occur for only an hour or so just before sunrise during the time they are flowering can result in severe losses. Failure to harvest sweet corn at the time it is at peak milk content results in overmature and tough sweet corn of low or no market value. Failure to apply insect control at a critical stage of the specific insect's life cycle results in ineffective control and severe crop loss. Such examples are infinite. Seldom does a single management decision lead to disaster or unbounded success, but being right "most of the time" is the highest sense of accomplishment a farmer can achieve.

What a farmer does is dictated by the physical and chemical composition of soil and land within the confines of the local economic, social, and political infrastructure, but *when* is determined by the individual artistic ability of the individual farmer. Within every farming community there are some farmers that are considered by their peers to be better than average. A reputation for conducting management operations on time invariably is associated with the better farmers in a community. The timing must be specifically applied to specific land and soil conditions in response to weather conditions.

Certainly the one scenario outlined by the corporate farm manager in the United States does not fairly represent all corporations. Large commercial farming operations are often criticized because of their competition with small farmers. Often large corporate farms are able to compete better because they have financial resources that enable them to adopt new technology quickly. Small farmers often do not have the financial capability to make investments in new machinery or soil-improving technology.

Many agricultural research programs are criticized as helping only the big farmers who are immediately able to adopt new technology. One newly graduated agricultural scientist who as a student had been severely critical of several research projects he thought only benefited large farmers was hired by an international corporation to manage their farming operations on a large but remote island in the Pacific Ocean. After a couple of years on the job, he wrote his former professors to explain that his opinion of their research and the role of big corporate farming had changed. He explained that because of the presence of the corporate farm on the island, the company was able to establish reliable markets both to buy farm products and market supplies to small farmers. Because of the corporate farm size, shiploads of farm supplies regularly docked in their harbor and departed with farm produce. With access to reliable markets for their crops and reliable supplies of seed and fertilizer, the small farmers had increased their production and were benefiting financially. I responded to the letter and in so doing pointed out some of the problems that often occur within large corporate farming operations, specifically mentioning that sometimes the business office overrules the farm manager. A humorous conclusion to this scenario occurred a few months later when I had the opportunity to visit the area in question. The former student flew some 500 miles to join me. Upon meeting me at the airport he asked, "How did you know the workings of my company so well?" I replied that I did not even remember the name of his company. He then explained that he had been planning to plant about 500 acres (202 ha) of cotton during the next growing season. To facilitate delivery of the necessary lime, fertilizer, seed, and pesticides, orders had to be placed six months in advance of planting. On the way to meet me he had stopped at corporate headquarters and inquired the status of his orders. He was informed that the lime was being unloaded that day, the seed had been delivered the night before, and the ship with the fertilizer would dock in two days. However, the bookkeeper informed him that his order for pesticides had been canceled because the company had a "temporary cash flow problem." He grimaced and said, "I have to figure out what I am going to do with the 500 acres of land and all that fertilizer, seed, and lime. "It is absolutely useless to plant cotton if no pesticides are available." Reliable *on time* conduct of farming operations are vital to obtaining high yields and marketing food products from the field. Timing is the component of the food supply that has to rely on the judgment of the individual farmer but in which all of society plays a role.

THE BOTTOM LINE

Throughout this book a brief overview of the physical, chemical, and biological functions of life were presented, made from the biased viewpoint of the human species. Regardless of various concepts of how the human species came into being, we are here. As a species of life we have the mental competence to command an exorbitant amount of energy and power. We have the ability to bend many other species to our will. We are mobile and thus able to cope with a vast array of climatic conditions. We have claimed portions of all kinds of land to grow and graze food for our survival as a species. We have tapped power from oil, coal, and minerals below the land surface. We have harnessed the air to support flight of airplanes and its movement to generate power. We harvest a small portion of the radiant energy from the sun via solar panels to generate power and heat some of our abodes.

Humans have the capability to outcompete other species, both plant and animal, to obtain what they need and/or want. All species that stand in our way will fall if human needs can be fulfilled by their extermination. The conversion of solar radiation via plants to nourish life is fundamental to all life. Ultimately, the number of people and how efficiently radiation is converted to human food will determine the amount of land claimed to harvest radiation for human food production via the solar panel of land.

A few years ago a cartoon characterization entitled "Natural System" came to my attention (Figure 10–10). For many years it seemed like a proper illustration of the

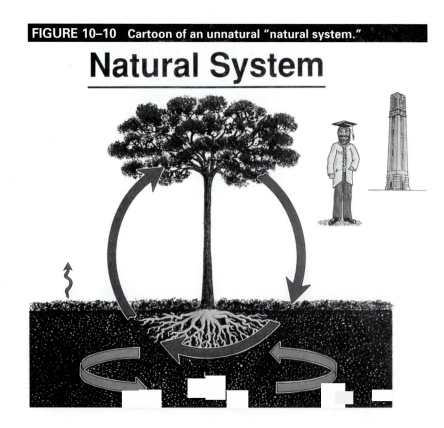

FIGURE 10–10 Cartoon of an unnatural "natural system."

Natural System

natural extraction of essential elements from the soil, their incorporation into organic compounds in a tree, and their return to the soil upon the death of the biological cells. After utilizing the figure in classes for several years, something seemed strangely out of place. I had been around university campuses most of my life and observed a great number of professors such as the one symbolically standing under the bell tower in the figure. I had never seen a professor with a broad smile on his face if he had not been fed. I must conclude that the cartoon is mislabeled; it really depicts an "unnatural system" in a world where humans exist.

Although all land surfaces support some form of life, only a small portion of the earth's total land has been found to be suitable for the production of human food crops. As a species, humans have found it necessary to displace and in some cases annihilate other species of life to claim land for their survival. As humans have learned to better understand the chemical and physical requirements of life, it is now possible to provide the food necessary for each human individual on less land than ever before. Where the advancements in agricultural technology have been applied, there has been a substantial reduction in the land area necessary to produce sufficient food for each human. However, there are now more humans than ever before, and the application of food-producing technology has not been uniformly applied throughout the world. From among the many technologies available to grow human food, selections must be made for compatibility with soil and land properties. Social and political climates dictate the application of agricultural technologies in much the same way as atmospheric climate and soil composition dictate the distribution of ecosystems on the land and tectonic movements and erosion shape landscapes. Prior to human habitation, energy from the sun coupled with the chemistry of minerals in soil naturally determined the portrait of life, that is, natural ecosystems, on the land. With increasing energy available to humans, and their determination to survive as a species, the future portrait of life on the land will continue to be altered as management practices necessary to produce human food are applied. The degree and extent of alteration inflicted by humans will depend on their numbers and the extent to which they efficiently utilize the land available to sustain human life.

LITERATURE CITED

Argabright, M. C., R. G. Cronshey, J. D. Helms, G. A. Pavelis, and H. R. Sinclair, Jr. 1996. *Historical Changes in Soil Erosion, 1930–1992: The Northern Mississippi Valley Loess Hills* (92 pp.). Historical Notes No. 5. Natural. Resources Conservation Service and Economic Research Service. Washington, DC: U.S. Department of Agriculture.

Helms, J. D., G. A. Pavelis, S. Argabright, R. G. Cronshey, and H. R. Sinclair, Jr. 1996. National Soil Conservation Policies: A Historical Case Study of the Driftless Area. *Agricultural History* 70:377–394.

Klingebiel, A. A., and P. H. Montgomery. 1961. *Land-Capability Classification.* Agriculture Handbook No. 210. Washington, DC: Soil Conservation Service, U.S. Department of Agriculture, U.S. Government Printing Office.

van Straaten, P. 2002. *Rocks for Crops: Agrominerals of Sub-Saharan Africa* (338 pp.). Nairobi, Kenya: ICRAF.

Wischmeier, W. H., and D. D. Smith. 1978. *Predicting Rainfall Erosion Losses: A Guide to Conservation Planning* (58 pp.). Agriculture Handbook No. 537. U.S. Department of Agriculture, Science and Education Administration.

CHAPTER REVIEW PROBLEMS

1. In addition to individual differences between land areas and soil types, what are some of the other factors that determine the practices that individual farmers employ?

2. Identify aspects of food production, from field to dinner plate, that farmers cannot control.

3. Identify the political and economic requirements that the entire society must assume to assure reliable human food.

4. Is environmental quality enhanced or degraded when human inhabitants of the land have an adequate supply of food? Defend your answer.

5. What do you suggest as policy(s) to assure reliable food production and environmental quality?

Appendix I

Soil Orders as Classified by the National Cooperative Soil Survey

For more detailed soil classification, see Soil Survey Staff, *Soil Taxonomy* (2nd ed.), Agriculture Handbook 436 (Washington, DC: USDA-Natural Resources Conservation Service, 1999). For sale by the Superintendent of Documents, U.S. Government Printing Office, Washington, DC 20402. Available online at http://soils.usdagov/technical/classification/taxonomy/.

Soil orders are the most general level of a hierarchical system of classifying soils that at present identifies more than 22,000 different kinds of soils in the United States and provides for identification of all soils in the world. Individual kinds of soils in the United States are known as Soil Series. The Official Series Description (OSD) of each Soil Series identified by National Cooperative Soil Survey soil-mapping projects in the United States is available at http://soils.usda.gov/technical/classification/osd/

index.html. There each Soil Series is scientifically classified according to Soil Taxonomy, the description of a typical pedon (soil at a specific site called a type location) is presented, and information regarding geographic setting, use, and vegetation, published references, and other information about each soil is presented. A very interesting website that interactively identifies where in the United States soils classified by each Soil Series name has been mapped by the National Cooperative Soil Survey is http//www.cei.psu.edu/soiltool/semtool_phase2. html.

An excellent reference that includes photos of representative soils and maps of their distribution in the United States and throughout the world is http://soils.ag.uidaho.edu/soilorders/. Educational material about soils in the United States can be accessed at http://soils.usda.gov/.

Order Name	Brief Description, Use, and Distribution
Alfisols (9.7%)[1]	Naturally forested soils of high natural fertility formed in nutrient-rich geologic material and extensively used around the world for food crop production. Major areas in the United States are in Ohio and surrounding areas. Extensive areas are also present in Europe and West Africa.
Andisols (0.7%)	Soils formed in volcanic ash. Many have high natural fertility and are intensively farmed. Major areas in Japan, Indonesia, Alaska, Central America, and Rift Valley of Africa.
Aridisols (12.0%)	Mineral rich soils in arid climates that must have irrigation for crop production. They are extensive in southwestern states in the United States, northern and southwestern Africa, Argentina, Central Australia, Central Asia, and the Middle East.
Entisols (16.2%)	Soils formed on steep slopes with very rapid natural erosion and in floodplains with very rapid deposition. Most floodplain locations are chemically fertile, and many are extensively used for crops. Locations on steep slopes are often woodland or range. Worldwide distribution in floodplains and mountainous areas.
Gelisols (8.6%)	Soils with permanently frozen subsoil. No crop production. Major areas in the Arctic regions of Siberia, Canada, and Alaska and in the Antarctic.
Histosols (1.2%)	Soils formed from organic residue and are naturally saturated with water. Artificial drainage systems are needed for crop production. Chemical fertility is usually high, but many are extremely acid. Worldwide distribution on level lowland with major areas in Finland, Russia, and Canada.
Inceptisols (9.8%)	Soils with abundant weatherable minerals on nonarid steep slopes with rapid natural erosion or on floodplains with rapid deposition. Chemical fertility is highly variable, related to mineral composition of geologic material. Extensive crop use worldwide where formed in nutrient-rich parent material.
Mollisols (6.9%)	Soils formed on chemically rich geologic material, in humid to semiarid climates under natural grass vegetation. They are intensively used for crop production. Major areas in the Midwest of the United States, Eastern Europe, and northern Argentina.
Oxisols (7.5%)	Soils formed on materials with few nutrient-rich minerals. Major areas in Brazil and Central Africa. Without external fertilization, no cropping or limited slash-and-burn farming is possible. With fertilization and modern technology, extensive areas are now producing excellent yields of food crops in Brazil.
Spodosols (2.6%)	Soils formed in sandy, chemically infertile geologic material under forest vegetation. Limited use for crops. Major areas are present in Florida and the northeastern United States, Canada, northern Europe, and Russia.
Ultisols (8.5%)	Soils formed on chemically poor rock and coastal plains under humid climates and forest vegetation. Without external sources of fertilizer, only slash-and-burn cropping is possible. Major areas are present in the southeastern United States, southern Asia, upper Amazon basin, and southern parts of Africa.
Vertisols (2.4%)	Soils formed from clay deposits or basic rock. They are chemically fertile but the shrink-swell characteristics of the clayey material present problems for cultivation. Many areas have limited rainfall. Major areas in Texas, Uruguay, Sudan, Australia, and India.

[1]Approximate percent of the world land area.

Appendix II

Conversion Factors for International Standard Units (SI)* and Non-SI Units, Fertilizer Oxide and Element Values, and Per Bushel Grain Weights

To Convert Column 1 to Column 2, Multiply by (Below)	Column 1 SI Unit (Abbreviation)	Column 2 Non-SI Unit (Abbreviation)	To Convert Column 2 to Column 1, Multiply by (Below)
		Units of Length	
0.621	Kilometer (km)	Mile (mi)	1.609
1.094	Meter (m)	Yard (yd)	0.914
3.28	Meter (m)	Foot (ft)	0.304
1.0	Micrometer (μm)	Micron (μ)	1.0
0.0394	Millimeter (mm)	Inch (in)	25.4
0.394	Centimeter (cm)	Inch (in)	2.54
		Area	
2.47	Hectare (ha)	Acre (Ac)	0.405
247	Square kilometer (km^2)	Acre (Ac)	0.00405
0.386	Square kilometer (km^2)	Square mile (mi^2)	2.590
10.76	Square meter (m^2)	Square foot (ft^2)	0.0929
0.00155	Square millimeter (mm^2)	Square inch (in^2)	645
		Volume	
35.3	Cubic meter (m^3)	Cubic foot (ft^3)	0.0283
6.10×10^4	Cubic meter (m^3)	Cubic inch (in^3)	1.64×10^{-5}
0.0284	Liter (L)	Bushel (bu)	35.24
1.057	Liter (L)	Quart (qt)	0.946
0.265	Liter (L)	Gallon	3.78
2.11	Liter (L)	Pint (pt)	0.473
		Mass	
0.0022	Gram (g)	Pound (lb)	454
0.0352	Gram (g)	Ounce (oz)	28.4
2.205	Kilogram (kg)	Pound (lb)	0.454
0.0011	Kilogram (kg)	Ton (2,000 lb)	907
		Density	
1.00	Megagram per cubic meter (mg m^{-3})	Gram per cubic centimeter (g cm^{-3})	1.00
624.6	Megagram per cubic meter (mg m^{-3})	Pounds per cubic foot (lb ft^{-3})	0.0016
168.6	Megagram per cubic meter (mg m^{-3})	Pounds per cubic yard (lb yd^{-3})	0.0059
		Yield and Rates	
0.893	Kilogram per hectare (kg ha^{-1})	Pound per acre (lb Ac^{-1})	1.12
0.0149	Kilogram per hectare (kg ha^{-1})	bushel per acre (60 lb bu)	67.19

0.0159	Kilogram per hectare (kg ha^{-1})	bushel per acre (56 lb bu)	62.71
0.0186	Kilogram per hectare (kg ha^{-1})	bushel per acre (48 lb bu)	53.75
893	Tons per hectare (T ha^{-1})	Pound per acre (lb Ac^{-1})	0.00112

Pressure

9.9	Megapascal (MPa)	Atmosphere (atm)	0.101
10	Megapascal (MPa)	Bar	0.1

Temperature

$(1.8 \times {}^\circ C) + 32$	Celsius (°C)	Fahrenheit (°F)	$0.5556 \times ({}^\circ F - 32)$

Quantity of Heat

0.239	Joule (J)	Calorie (cal)	4.19

Electrical Conductivity

10	Siemen per meter (S m^{-1})	Millimho per centimeter (mmho cm^{-1})	0.1

Concentrations

1	Centimole per kilogram (cmol kg^{-1})	Milliequivalents per 100 grams (mEq 100^{-1})	1
0.1	Gram per kilogram (g kg^{-1})	Percentage (%)	10
1	Milligram per kilogram (mg kg^{-1})	Parts per million (ppm)	1

Plant Nutrient and Oxide

	Element	Oxide (Numbers on a Fertilizer Bag)	
1	N	N (first number)	1
2.29	P	P$_2$O$_5$ (second number)	0.437
1.2	K	K$_2$O (third number)	0.83

*SI, Système Internationale (French).

Weight per Bushel of Common Grains

Type of Grain	Pounds per Bushel
Barley	48
Corn (shelled)	56
Corn (whole ear)	70
Oats	32
Rice	45
Sorghum	56
Soybeans	60
Wheat	60

Glossary[1]

Acid soil   Soil with a pH value less than 7.

Agribusiness   All businesses that supply material to farmers or buy, transport, or market agricultural products.

Agroforestry   Multiple cropping land use that combines growing of food crops and trees on the same land in mixtures or in sequence.

Agronomy   The science and practice of growing food crops and soil management.

Air dry soil   Water content in equilibrium with the humidity of the surrounding air.

Albedo   The fraction of light reflected from a surface.

Algae   Lower forms of plant life capable of *photosynthesis*.

Alkali soil   Soil with a pH value of 8.5 or higher; contains sufficient sodium to hinder the growth of most plants.

Alkaline soil   Soil with a pH value greater than 7.0.

Allophane   An aluminosilicate mineral with primarily short-range structural order occurring as exceedingly small spherical particles, especially in soils formed from volcanic ash.

Alluvial soil   Soil formed in recent material deposited by flowing water.

Amorphous material   Noncrystalline constituents that do not fit all of the criteria of *allophane*.

Anaerobic   Absence of molecular oxygen (O_2).

Anion   A negatively charged ion.

Anther   Part of a plant that contains pollen.

Aquatic conditions   Continuous or periodic saturation with water so as to limit the presence of molecular oxygen in the soil.

Aquifer   A saturated, permeable geologic layer of sediment or rock that can transmit significant quantities of water under hydraulic gradients.

Arable land   Land suitable for the production of food crops.

Arroyo   A term used in the arid and semiarid regions of the southwestern United States. Small deep channels or gullies of an ephemeral stream, usually with vertical or steeply cut banks of unconsolidated material.

Aspect   Compass direction toward which a slope faces (e.g., north facing, east facing, etc.).

Available nutrients   Amount of an essential element in the soil in a chemical form that can be accessed by plant roots during the growing period of that plant.

Available water-holding capacity   Amount of water released between in situ *field moisture capacity* and the *permanent wilting point*.

Azotobacter   Flagellated, nonsymbiotic bacteria in soil.

Backswamp   Marshy or swampy depressed areas of floodplains between natural levees and valley slopes.

---

[1] Primary sources: *A Geomorphic Classification System*. 1998. Washington, DC: USDA Forest Service; *Glossary of Crop Science Terms*. 1992. Madison, WI: Crop Science Society of America; *Glossary of Soil Science Terms*. 1997. Madison, WI: Soil Science Society of America.

Banding Practice of placing fertilizer in a strip next to a planted crop.

Bars A ridgelike accumulation of sand, gravel, or other alluvial material formed in the channel of a stream where a decrease in velocity induces deposition.

Beach Area of land, usually sandy and devoid of vegetation, between open water and vegetated land.

Bedrock Solid rock, either under soil or exposed on the land surface.

Biodegradable A material that can be decomposed by natural processes.

Biomass Total weight of all organisms in a given area.

Bioremediation Use of biological agents to reclaim contaminated soil.

Biotite A mica mineral containing iron and magnesium; flat brown flakes.

Bottomland Land at the base of a hill, often subject to flooding (see *Floodplain*).

Broadcast Practice of applying fertilizer or other material to the surface of the soil.

Bulk density Mass weight of soil per unit volume of soil: Bulk density = weight/volume. In the metric system, bulk density is calculated as the weight of soil in grams divided by the volume of the soil sample in cm^3 (BD = g cm^{-3} or mg m^{-3}; see Appendix).

Bypass flow See *Preferential flow*.

Calcareous soil Soil containing enough free $CaCO_3$ and other carbonates to effervesce visibly when cold 0.1 molar HCl is applied.

Calorie The amount of heat required to raise the temperature of 1 g of water 1°C.

Capillary water Water held in the small voids (approximately less than 0.01 mm in diameter) that does not move downward under the force of gravity. (Term is no longer used by soil scientists to describe slow upward movement of water into voids above a *water table*.)

Carbon-to-nitrogen ratio (C:N ratio) The ratio of the mass of organic carbon to the mass of organic nitrogen in soil and organic materials.

Cat clay Poorly drained clay textured soil containing ferrous sulfide that becomes very acid if drained and allowed to oxidize.

Catena Spatial association of soils with different depths to a water table. Also called a toposequence (see Chapter 7, "The Hill Family").

Cation A positively charged ion.

Cation exchange Interchange of positively charged ions between negatively charged clay or organic matter surface and the surrounding soil water.

Cation exchange capacity (CEC) The amount of negative charges in a soil capable of attracting positively charged ions. (Various methods are used because CEC differs with pH of the method.) Measurements expressed as $cmol_c$ kg^{-1} of soil, or in older literature as milliequivalents (mEq) $100g^{-1}$ of soil.

Chaff Glumes, hulls, and small fragments of straw separated from grain or seed in the threshing process.

Chemical weathering Breakdown of rocks and minerals resulting from the presence of water and chemical substances in the water.

Chlorophyll Green pigment in plants necessary to the process of *photosynthesis*.

Chroma Relative purity of a color. Low chroma values are associated with grayness.

Clay Mineral particles less than 0.002 mm in equivalent diameter. Most clay is a *secondary mineral*.

Climax Stage in *ecosystem* development in which a community of organisms becomes stable and begins to perpetuate itself.

Coliform General term for bacteria present in the intestinal tract of humans and other animals.

Colluvium Unconsolidated earth material deposited as a result of a *landslide*.

Color (soil) Soil colors are specifically identified by the Munsell color system that specifies three variables: *chroma, hue,* and *value* (e.g., 7.5 YR 6/4 where 7.5 YR is hue, 6 is value, and 4 is chroma).

Combine Machine that cuts grain crops and separates the seed grain from the stems, *chaff*, and leaves while moving across a field. The seed grain is retained and hauled away; the stems, chaff, and leaves are distributed over the land.

Compaction A decrease in the volume of soil material resulting from applied pressure. It results in an increase in *bulk density*.

Compost   Organic residues or a mixture of organic residues and soil that have been mixed, piled, and moistened with or without addition of fertilizer and lime and allowed to undergo decomposition until the original organic materials have been substantially altered or decomposed.

Composting   A biological process in which organic residues, usually wastes, are piled in containers and allowed to slowly decompose to *humus*-like material suitable for use as a soil amendment or organic fertilizer.

Contour strip cropping   Growing of two or more crops in alternating strips of land perpendicular to the downward flow of water on sloping land.

Cover crop   Vegetation planted for the purpose of protecting the soil from erosion during the time regular crops are not growing and/or between widely space crops as in orchards.

Creep   Slow mass movement of soil material down slope.

Crest   Top of a hill, ridge, or mountain.

Crust   A transient soil surface layer ranging in thickness from a few millimeters to a few centimeters that is either denser or more cemented than the material immediately below.

Crystalline rock   Rocks composed of various minerals that formed at high temperatures from magma.

Cultivar   A variety, strain, or race of plant that has originated and persisted under cultivation or was specifically developed for the purpose of cultivation.

Cultivate   See *Tillage; cultivation.*

Cultivation   Shallow *tillage* operation conducted to create soil conditions more conducive to crop growth or to control weeds.

Cytoplasm   Cellular material within the cell membrane and surrounding the nucleus.

Decalcification   Removal of calcium carbonate and calcium ions by leaching.

Delta   Nearly flat area of land at or near the mouth of a river where that river enters a lake or ocean.

Denitrification   Chemical reduction of nitrogen oxides (usually NO_3^-) to molecular nitrogen (N_2) or nitrogen oxides with a lower oxidation state.

Depression   A low-lying area in the land surface surrounded by higher elevations.

Dissection   Formation of gullies or valleys surrounded by hills or flat-topped remnants between the river or stream channels resulting from water erosion of a land surface.

Distilled water   Water prepared by evaporation and condensation processes to remove all soluble salt.

Divide   Area of higher ground that forms the boundary between two adjacent drainage basins.

Dolomitic lime   Liming material composed of magnesium and calcium carbonates ($Ca:MgCO_3$).

Drain tile   A rigid but porous pipe buried in the soil to remove water from a field.

Drainage, engineered   Any engineered system installed in a field to lower the *water table* or otherwise remove excess water.

Dry-land farming   Crop production without irrigation.

Dune   A low mound of loose windblown material, usually sand.

Dust mulch   A loose or powdery condition of the soil surface.

Ecosystem   Energy-driven complex of one or more organisms and their environment.

Eolian   Soil material transported and deposited by wind (e.g., sand dunes, *loess*).

Erosion   The removal of material from the surface of the land by wind, water, or ice.

  Accelerated erosion   Erosion in excess of natural rates resulting from human activity.

  Geological erosion   Normal or *natural erosion* caused by geological processes.

  Gully erosion   Formation of deep channels, from 20 inches to 100 feet deep, resulting from water flowing down sloping land.

  Interrill erosion   Nearly uniform removal of surface material by splashing due to the impact of raindrops.

  Natural erosion   Erosion not related to human activity (see *Geological erosion*).

  Rill erosion   Numerous small channels due to flowing water on side slopes.

  Sheet erosion   Erosion of a relatively uniform thin layer of soil by raindrop impact and largely unchanneled surface runoff.

Essential elements Chemical elements required by plants (C, H, O, P, K, N, S, Ca, Fe, Mg, Mn, Cu, B, Zn, Co, Mo, Cl, and Na).

Evapotranspiration Loss of water to the air from evaporation from the soil surface and transpiration from plants.

Actual evapotranspiration The actual amount of water loss from the soil by evaporation and transpiration considering little or no loss when there is not a complete cover of actively growing vegetation.

Potential evapotranspiration The potential for water loss as calculated from air temperature and day length when the soil is vegetated with a complete canopy of actively growing vegetation that is well supplied with water.

Exchangeable bases Ca^{2+}, Mg^{2+}, K^+, and Na^+ ions attracted to the negative charges on clays and organic matter.

Exchangeable sodium percentage (ESP) Amount of exchangeable sodium as a percentage of total *cation exchange capacity*. ESP = ($cmol_c$ Na^+ kg^{-1}/$cmol_c$ CEC kg^{-1} soil) × 100).

Extractable soil nutrient Quantity of a nutrient removed from a soil sample by a specific soil-testing procedure.

Fallow period Time when a normally cultivated area of land is not planted to a food crop but allowed to remain barren or natural vegetation is allowed to grow.

Fan Landform at the base of a slope where eroded material from that slope is deposited (see *Foot slope* and *Toe slope*).

Fertility (soil) Relative ability of a soil to supply the nutrients essential to plant growth.

Fertilizer An organic or inorganic material of natural or synthetic origin (not including *lime*) added to soil to provide one or more of the essential nutrients.

Blended fertilizer Mechanical mixture to different fertilizer materials.

Complete fertilizer Contains nitrogen, phosphorus, and potassium and often other essential nutrients.

Controlled-release fertilizer Fertilizer material coated so as to delay dissolution in the soil. Also called slow-release or slow-acting fertilizer.

Inorganic fertilizer Fertilizer in which carbon is not an essential component of its chemical structure.

Liquid fertilizer Liquid material containing dissolved and/or suspended components.

Organic fertilizer Material containing carbon and one or more plant nutrients.

Side-dressed fertilizer Fertilizer applied adjacent to a planted crop after that crop has emerged from the soil.

Starter fertilizer Fertilizer applied at the time of planting to speed early growth.

Top-dressed fertilizer Fertilizer applied to the surface after the crop is growing.

Fertilizer analysis Chemical composition expressed as a percentage of total nitrogen, available phosphoric acid (P_2O_5), and water-soluble potash (K_2O), (P_2O_5 contains 44% P; K_2O contains 83% K). In most areas of the world, the composition of fertilizer must be printed on the container and is guaranteed by law. Fertilizing material labeled "soil amendments" usually have no guaranteed analysis or less nitrogen, phosphorus, and potassium than the legal definition of fertilizer.

Field moisture capacity The content of water in the soil two or three days after the soil has been wetted and allowed to drain freely.

Fifteen atmosphere (or bar) water Laboratory approximation of the water remaining in the soil when most plants wilt (see *Permanent wilting point*).

Fixation Process(s) by which plant nutrients in the soil are rendered less or unavailable to plants.

Floodplain Nearly level area of land that borders a river or stream and is subjected to inundation during flood events.

Flower bud Contains one or more embryonic flowers.

Foot slope Gently inclined surface at the base of a slope (see *Toe slope*).

Forage Edible parts of plants, other than separated grain, that can be harvested for feeding animals.

Forage crop A crop of cultivated plants produced to be grazed harvested for use as feed for animals.

Forest floor Organic matter generated by forest vegetation including *litter* (*organic* *residue*) and unincorporated humus resting above the mineral soil surface.

Forest grazing Combined use of forestland and woodland for both wood production and animal grazing.

Forestland Land on which the vegetation is dominated by forest.

Frost-free days Most often reported as the *average* number of days between the last minimum temperature of $32°F$ ($0°C$) in the spring and the first return of freezing in the fall.

Gene The physical and functional unit of heredity.

Gene bank Collection of cloned DNA fragments that, ideally, represent all the sequences of a single *genome*.

Genetic engineering The use of in vitro techniques to produce DNA molecules containing novel combinations of *genes* or other sequences of living cells that make them capable of producing new substances or performing new functions.

Genome The hereditary material of a cell found in each *nucleus* or organelle of a given species.

Geomorphic surface Land surface that has a common geologic history of age and surface shape and is formed by processes of landscape evolution.

Germination Resumption of active growth by the seed embryo, culminating in the development of a young plant.

Germplasm The living substance of the cell nucleus that determines the hereditary properties of the organism and transmits these properties to the next generation.

Gilgai Microrelief of small basins and knolls or valleys and ridges on a soil surface produced by expansion and contraction during wetting and drying of clayey soils that contain *smectite*.

Glacial drift General terms for all mineral material transported and deposited from glaciers.

Gley soil (Gleyed) Condition resulting from prolonged saturation during which iron oxides have been reduced, thus leaving a gray (usually 2 or less *chroma*) colored soil.

Glomalin Protein-sugar substance secreted by certain fungi that helps soil particles form soft aggregates.

Grassed waterway A natural or constructed waterway covered with grass vegetation and not tilled, placed so as to conduct surface water flow down slope through cultivated fields with minimum erosion.

Grassland Land on which the vegetation is dominated by grasses.

Gravitational water Water that moves into, through, or out of the soil under the influence of gravity.

Grazing land Any vegetated land that is grazed or has the potential to be grazed by animals.

Green manure Plant materials mechanically incorporated into the soil while still growing or shortly after the plants have died.

Green manure crop Any crop grown for the purpose of being incorporated into the soil while still growing or shortly after maturity.

Groundwater Water saturated zone in and/or below the soil and under the *water table*.

Guano Decomposed and dried excrement of birds and bats used for fertilizer.

Gullied land Areas where all the natural *soil horizons* have been removed by water erosion.

Gully A water-eroded channel more than about 2 feet deep that interferes with normal *tillage* operations.

Hardwood Wood produced by broad-leaved trees such as maple, oak, and elm.

Heavy metals Metals that have densities more than 5.0 mg m^{-3}. (In soil these include Cd, Co, Cr, Cu, Fe, Hg, Mn, Mo, Hi, Pb, and Zn.)

Hue Chromatic composition (*color*) of light reflected from an object.

Humus Partially decomposed organic material. Often used synonymously with *soil organic matter*.

Hybrid An organism resulting from the controlled cross-fertilization between individuals differing in one or more genes.

Hydric soils Soils that are wet long enough to periodically produce *anaerobic* conditions that influence the growth of plants (regulatory definitions used in defining wetlands).

Hydrophytic vegetation Plants growing in water or on a soil that is at least periodically deficient in oxygen as a result of excessive water content.

Hygroscopic water Water remaining in the soil after air drying.

Igneous rock Rock formed from the cooling of magma.

Illite General term used for *mica*.

Immobilization The conversion of an element from the inorganic to the organic form in microbial or plant tissues.

Impeded drainage A condition that slows the movement of water downward in soils.

Indigenous Native to an area.

Infiltration Entry of water through the soil surface.

Intercropping Planting two or more crop species within the same area at the same time.

Interflow The lateral down-slope movement of water that has *infiltrated* and is below the soil surface.

Interfluve Landform composed of relatively undissected upland or a ridge *crest* between two adjacent valleys or drainage ways.

Ions Atoms, groups of atoms, or compounds that are electrically charged as a result of the loss of electrons (cations) or the gain of electrons (anions).

Irrigation Intentional application of water to the soil. See the following types:

Center-pivot irrigation Automated sprinkler system that rotates a boom around a pivot source of water. The circle formed by a typical pivot covers 130 acres (52.7 ha).

Drip irrigation Water is slowly applied to the soil surface through small emitters having low-discharge orifices.

Flood irrigation Water is released from field ditches and allowed to flood over the land.

Furrow irrigation Water is applied between crop rows in shallow ditches created by tillage implements.

Sprinkler irrigation Water is sprayed over the land surface through nozzles or high-volume guns utilizing a pressurized system.

Subirrigation Water is applied in open ditches or tile lines until the water table is raised to the rooting depth of the crop.

Subsurface drip irrigation Water is applied below the soil surface through small emitters.

Kaolin Subgroup name for aluminum silicates with a 1:1 structure.

Kaolinite Most common clay mineral in the *kaolin* subgroup.

Karst Topography with sinkholes, caves, and underground drainage that is formed by dissolution in limestone.

Killing frost Freezing condition that kills the plant. The exact temperature necessary to kill a plant differs among species and stage of plant growth. Most often a killing frost is several degrees below $32°F$ ($0°C$).

Lacustrine Formed (deposited) in a lake.

Land capability classification Classification of land as regards its suitability for human uses and potential for degradation (see Chapter 10).

Landform Any physical form or feature on the earth's surface having a characteristic shape and produced by natural processes.

Landscape A collection of related *landforms* that can be seen from a single viewing point.

Landslide The rapid down-slope movement under the influence of gravity of a mass of soil, rock, and other earthen materials.

Law of the minimum States that the growth and reproduction of an organism depends on the nutrient substance that is available in minimum quantity.

Leaching Downward removal of water or soluble materials from one layer in soil to another or completely from the entire soil depth.

Legumes Members of the Leguminoseae, the pea or bean family.

Levee, natural A low ridge or embankment of sand and coarse silt built by a stream on its *floodplain* in time of flood when water overflowing the normal banks is forced to deposit the coarsest part of its load.

Lime, agricultural Soil amendment containing calcium carbonate, magnesium carbonate, and other material. Used to neutralize soil acidity and furnish calcium and magnesium for plant growth.

Litter Surface layer of leaves, needles, and branches on the forest floor that are not in an advanced stage of decomposition.

Loess Wind-deposited material, mainly of silt-sized particles.

Lysimeter   A device placed under an area of soil to measure the gains and losses of moisture, usually by weighing the soil above and/or collecting water that leaches through the soil.

Macronutrients   General term for those plant-essential elements required in relatively large amounts. Usually refers to nitrogen, phosphorus, and potassium but may include calcium, magnesium, and sulphur.

Manure   Excreta of animals, with or without admixtures of bedding or litter; fresh or at various stages of further decomposition or composting.

Marl   Soft calcium carbonate usually mixed with varying amounts of clay or other impurities.

Meander belt   A zone along a valley floor across which a meandering stream shifts its channel from time to time.

Mica   A layered aluminosilicate mineral group of the 2:1 type containing potassium.

Microbe   A very minute organism (i.e., a microorganism). Microbes include bacteria, protozoa, fungi, and so on.

Microclimate   Climatic conditions very near the land surface in a very small area of land resulting from local differences in elevation or exposure (*aspect*).

Micronutrient   Plant-essential nutrient found in relatively small amounts in plants (usually B, Cl, Cu, Fe, Mn, Mo, Ni, Co, and Zn).

Microrelief   Slight irregularity in form and height of a land surface that is superimposed on a larger *landform* such as low mounds, swales, and shallow pits.

Mineral   Naturally occurring inorganic solid particle with a definite chemical composition and an ordered atomic arrangement.

Mineralization   Conversion of an elemental from an organic form to an inorganic state as a result of microbial activity.

Monoculture   Growing of one crop species on an area of land each growing season.

Montmorillonite   Clay mineral with a 2:1 layered structure of two silica tetrahedral sheets and a shared aluminum and magnesium octahedral sheet. Montmorillonite greatly expands when wet and shrinks when dry.

Moraine   Mound or hill of material deposited by a glacier.

Mottles (soil color)   Spots or blotches of different color interspersed within the dominant color of the soil.

Muck, soil   Organic soil material so decomposed that the plant parts cannot be identified.

Mucky peat, soil   Organic soil material partially decomposed with some plant parts identifiable.

Mulch   Any material such as straw, sawdust, leaves, plastic film, loose soil, and so on, spread on the surface of the soil to protect the soil and/or plant roots from the effects of raindrops, freezing, evaporation, and so on.

Muscovite   Clear silicate mineral in the *mica* group.

Mycorrhiza (Mycorrhizae, *pl.*)   Specific fungi that form a symbiotic relationship with roots of higher plants and act as an extended root system of that plant.

Neutral soil   Surface layer of soil with a pH value between 6.6 and 7.3.

Nitrogen-fixing bacteria (microbes)   Bacteria living in the soil or in the roots of *legumes* that convert atmospheric nitrogen (N_2) into nitrogen compounds in their own bodies.

No-Till   Procedure whereby a crop is planted directly into the soil with no primary or secondary *tillage* since harvest of the previous crop.

Nucleus   Body within a living cell that contains the cell's hereditary material and controls its metabolism, growth, and reproduction.

Nutrient   Elements or compounds essential as raw materials for organism growth and development.

Nutrient balance   Undefined theoretical ratio of two or more plant *nutrient* concentrations for an optimum growth rate and yield.

Nutrient deficiency   Low concentration of an essential element that reduces plant growth and/or prevents completion of the normal plant life cycle.

Nutrient stress   Condition when the quantity of *nutrients* available reduces growth.

Nutrient toxicity   Harmful effect from a high essential *nutrient* concentration in a plant.

Oligotrophic   Environments (soils) in which the concentration of **nutrients** available for plant growth is limited.

Organic farming Crop production system that reduces, avoids, or largely excludes the use of synthetically compound fertilizers, pesticides, growth regulators, and livestock feed additives.

Organic fertilizer Product made from processing of animals or vegetable substances that contains sufficient plant nutrients to be of value as fertilizer.

Organic residue Plant and animal remains in the early stages of decomposition; enough of the original structure remains that its source can be identified.

Organic soil See Histosols, Appendix I.

Organic soil material Soil material containing 180 g kg^{-1} (18% by weight) or more organic carbon if the mineral fraction has 600 g kg^{-1} clay, or 120 g kg^{-1} (12% by weight) or more organic carbon if the mineral fraction has no clay or has proportional intermediate contents of organic carbon.

Osmosis The net movement of a solution through a living cell membrane from a solution of low dissolved salt concentration to a solution of high salt concentration.

Outwash Mainly stratified layers of sand and gravel deposited by water flowing from melting glaciers.

Oven-dry soil Soil dried at 105°C until it reaches a constant mass.

Oxidation Loss of one or more electrons by an ion or molecule.

Paddy A field where a crop, most frequently rice, is flooded during part or all of the growing season.

Paleosol Soil formed on a landscape in the past with distinctive morphological features resulting from a soil-forming environment that no longer exists at the site.

Pan A layer in the subsoil that slows or prevents water movement and/or plant root growth.

Parent material Geologic material from which a soil has formed (also called initial material).

Particle-size distribution Proportions of sand, silt, and clay in a soil sample. Often called *soil texture* (see Chapter 2).

Parts per million (ppm) Concentration in a solution expressed in weight of a dissolved substance per million weight of the solution. Numerically equal to milligrams per kilogram (mg kg^{-1}).

Pastureland Land devoted to the production of indigenous or introduced forage for harvest primarily by grazing domestic animals.

Peat Slightly decomposed *organic soil material* in which the original plant parts can be recognized.

Peatland Generic term for any *wetland* that accumulates partially decayed plant matter.

Perennial, plant Persisting for several years, usually with new growth from a root.

Permafrost Permanently frozen layer in a soil.

Permanent wilting point Soil moisture content at which most plants die from lack of water (see Chapter 2).

Permeability Ease with which gases, liquids, or plant roots penetrate or pass through a mass or layer of soil.

pH, soil The pH value of a solution in contact with the soil material.

Phloem Conducting tissue present in *vascular plants*, chiefly concerned with the transport of food materials in the plant.

Phosphate Fertilizer trade terminology for soluble phosphoric acid (P_2O_5) and often referred to as available phosphoric acid (P_2O_5 contains 44% elemental P).

Phosphorus, fixation Immobilization of phosphorus by strong adsorption or precipitation such that it is not available to plants.

Photosynthesis Manufacture of carbohydrate from carbon dioxide and water in the presence of *chlorophyll* using light energy and releasing oxygen: $CO_2 + 2H_2O + $ light energy $\rightarrow (CH_2O) + O_2 + H_2O$.

Physical weathering Breakdown of rock and mineral particles into smaller particles by physical forces such as frost action.

Plaggen Human-made layer of soil more than 20 inches (50 cm) thick that is formed by long-continued application of *manure* and mixing.

Plain A nearly level area of land.

Plant food A common expression in the fertilizer trade for inorganic *nutrients* used by plants.

Plow pan *Pan* created by compacting soil at the lower boundary of a *moldboard* plow or other *tillage* implement.

Pollination   Transfer of pollen from the *anther* to the *stigma* of a flower.

Potash   Term used to refer to potassium fertilizers usually designating K_2O (K_2O contains 83% K).

Potassium fixation   Process whereby K^+ ions enter into certain clay structures and become unavailable to plants.

Pothole   Shallow marshlike pond.

Preferential flow   Rapid flow of water through large cracks or channels in soil without wetting entire soil mass.

Primary mineral   Mineral that have not been altered chemically since deposition and crystallization from molten magma.

Profile, soil   Vertical section of soil exposing all *soil horizons*.

Puddle, soil   Compaction of the soil to make it less *permeable* to the flow of water.

Pyroclastics   General term for volcanic materials aerially ejected from a volcanic vent.

Reaction, soil   General term for degree of acidity or alkalinity (i.e., *pH*).

Redox   Contraction for *reduction-oxidation* (e.g., redox reactions).

Reduction   Gain of one or more electrons by an ion or molecule.

Regolith   All loose earth materials above hard rock; includes soil.

Relief   Relative difference in elevation between the *crest* of the hills and valley *bottomland* within a *landscape*.

Residual fertility   Amount of available *nutrient* applied to one crop that remains in the soil after that crop is harvested and is available for subsequent crops.

Respiration   Oxidation of food and the release of energy, which may be either aerobic or anaerobic.

Rhizobia   Bacteria able to live symbiotically (*symbiosis*) in roots of leguminous plants from which they receive energy and in turn capture molecular nitrogen (N_2) from the air that can be used by the host legume plant.

Riparian   Land adjacent to a body of water that at least periodically is influenced by flooding.

Rock outcrop   An exposure of hard rock at the land surface.

Runoff   That portion of precipitation or irrigation water that does not *infiltrate* but instead flows over the soil surface away from the site.

Run-on   Runoff water that *infiltrates* another soil down slope from the site of *runoff*.

Saline soil   Soil containing enough soluble salt to affect the growth of most crop plants adversely. The conventional value of defining a saline soil is a value of 4 dS m^{-1} (at 25°C) or greater electrical conductivity in a *saturation extract*.

Saprolite   Soft, friable isovolumetrically weathered bedrock that retains the fabric and structure of parent rock.

Saturate   To fill all the soil voids with water or other liquid.

Saturation extract   Water or other liquid extracted from a soil that is saturated.

Savanna   Grassland with scattered trees or shrubs: often a transitional type between true *grassland* and *forestland*.

Secondary mineral   A mineral that forms the decomposition of a *primary mineral* or from the precipitation of the products derived from the decomposition of a primary mineral, usually a *clay*-size mineral.

Sediment   Transported and deposited particles derived from rocks, soil, or biological materials.

Sedimentary rock   Rock formed from materials deposited from suspension or precipitated from solution (e.g., sandstone, limestone, shale, etc.).

Shoulder   The hill slope position below the crest and the side slope (see Chapter 7).

Side slope   Usually linear slope below the *shoulder*. Also called back slope.

Silage   *Forage* preserved in a succulent condition by partial anaerobic, acid fermentation.

Sinkhole   A closed depression formed by dissolution of bedrock, usually *limestone*, and the collapse of the overlying material.

SI units   [Système Internationale (French)] International Standard units (see Appendix II).

Slick spots   Areas of land having a puddled or crusted nearly impervious soil surface.

Slow-release fertilizer   See *Fertilizer, controlled release*.

Slump   See *Landslide*.

Smectite   A group of 2:1 silicate minerals that includes *montmorillonite*.

Sodic soil See *Alkali soil.*

Sodium absorption ratio (SAR) The relation between soluble sodium and soluble divalent cations that can be use to predict the *exchangeable* sodium fraction of soil equilibrated with a given solution. SAR= $Na^+/(Ca^{2+} + Mg^{2+})^{1/2}$ where all units are expresses in mmoles per liter.

Softwood Wood produced by coniferous trees such as pine.

Soil The unconsolidated mineral and/or organic material on the immediate surface of the earth that serves as a natural medium for the growth of land plants.

Soil air The soil atmosphere in space not occupied by solids or water.

Soil compaction Increasing the *bulk density* by the application of mechanical forces to the soil.

Soil conservation Protecting soil against physical loss by erosion or against chemical deterioration that causes loss of fertility by either natural or artificial means.

Soil horizon(s) A layer in the soil, approximately parallel to the land surface, that differs in *color, texture,* or *structure* from the layers above and below. The main *soil horizon* identification codes (used in writing soil profile descriptions) are as follows:

O horizons Layers dominated (more than 20%) by *organic residue.*

A horizons Mineral layers with appreciable amounts of organic carbon, usually dark colored.

E horizons Light colored layers, usually below A horizons and above B horizons that have lost iron, clay, and/or organic carbon.

B horizons Mineral horizons usually formed below A horizons and sometimes E horizons if present where clay, iron oxide, carbonates, aluminum, or organic compounds have accumulated.

C horizons Horizons or layers of friable material below any or all of the preceding horizons.

R layers Hard bedrock of granite, limestone, basalt, sandstone, or other rock type.

Soil management All *tillage* and planting operations, cropping practices, fertilizer, lime, irrigation, herbicide, and insecticide (pesticides) application and other treatments conducted on or applied to a soil for the production of plants.

Soil micromorphology The study of *soil morphology* by microscopic methods.

Soil monolith A vertical section of a soil *profile* removed from the soil and mounted for display.

Soil morphology Visible characteristics of a soil or parts of a soil.

Soil sample A representative sample removed from a soil and transported for further study.

Soil Series The lowest category of the U.S. Soil Taxonomy soil classification system. Over 22,000 Soil Series are recognized and used to name soils on published *soil surveys* in the United States.

Soil structure Arrangement of soil particles into aggregates.

Soil survey The systematic examination, description, classification, and mapping of soils.

Soil Taxonomy System for classification developed by the National Cooperative Soil Survey program in the United States (see Appendix I).

Soil test Chemical, physical, or biological procedure that estimates the suitability of a soil to support plant growth.

Soil test interpretation The process of developing nutrient application recommendations from the soil test concentrations and other soil, crop, economic, environmental, and climatic information.

Soil texture *Particle-size distribution* (see Chapter 2).

Stand Number of established plants per unit land area (e.g., 27,000 corn plants per acre, etc.).

Steady state A condition in which the rate of input of a material equals the rate at which that material is removed. Thus the content of the material does not change over time.

Stigma Part of plant flower that receives pollen and on which they germinate.

Stocking rate The number of animals grazing an area over a given period of time.

Stomate (Stomata, *pl.*): Openings or pores between two specialized epidermal cells, the guard cells on the exterior of green plant tissues.

Stover Matured and cured stalks of such crops as corn and sorghum from which the grain has been removed.

Straw Matured stalks of grain such as wheat or oats from which the grain has been removed.

Subsoil General term most often referring to the *B horizon* but may include all soil material below the *A horizon*.

Succession The gradual and orderly process of *ecosystem* development brought about by changes in species populations and culminating in the production of *climax* characteristics of a particular geographical region.

Symbiosis Two or more organisms living together in an association that is mutually advantageous.

Tectonic Relating to deformation of the earth's crusts (i.e., faulting and folding).

Terrace, engineered A generally horizontal strip of earth and/rock constructed along a hill on or nearly on a contour to make land suitable for *tillage* and to reduce *accelerated erosion*.

Terrace, natural A steplike land surface bordering a steam or shoreline that represents the former position of a floodplain, lake, or seashore.

Tile drain Porous pipe or a related structure placed at a suitable depth and spacing in the soil to enhance and/or accelerate removal of water from the soil profile.

Till Unsorted earth material that has been deposited by glacial ice.

Tillage The mechanical manipulation of the soil profile for any purpose, but in farming it is usually restricted to modifying soil conditions to dispose of crop residue, weed control, incorporating chemicals, and/or creating a desirable condition for crop plants to germinate and grow. The following are some common types of tillage operations conducted by farmers:

Bedding Process of preparing a series of parallel ridges separated by shallow furrows.

Bed planting Seeds are planted on slightly raised areas (see *Bedding*).

Broadcast Uniform planting of seeds over the entire planted area.

Chisel To break up the soil using closely spaced gangs of narrow shank-mounted tools.

Clean tillage Burying all plant residues below the soil surface.

Conservation tillage Leaving 30% or more of the surface covered by plant residue.

Conventional tillage Leaving less than 30% of the surface covered by plant residue.

Cultivation (weeding) Shallow tillage to destroy weeds between plant rows of crop.

Drag To pull heavy and ridged implements across the soil surface to crush clods.

Minimum tillage Using minimum amount of tillage to provide good crop growth.

Moldboard plowing Partial or complete inversion to a depth of about 7 inches.

No-till Planting without tilling residues of previous crop. Use of chemicals to kill weeds.

Reduced tillage A tillage system in which the total number of tillage operations prior to planting is reduced from that normally used.

Strip cropping Growing two or more crops in alternate bands on the contour of the slope to reduce water erosion or perpendicular to prevailing wind direction to reduce wind erosion.

Subsoiling Loosening the soil below the surface horizon with minimum surface disruption.

Toe slope A gently inclined landscape surface at the base of a slope. Toe slopes have less of an incline than *foot slopes*, and in many instances the two terms are interchangeable.

Top dressing Application of fertilizer to the soil surface without any tillage.

Toposequence A sequence of related soils that differ from each other primarily because of their position on the landscape (see Chapter 7, "The Hill Family").

Topsoil Presumably fertile soil material removed from the surface layer of soil and applied to gardens, lawns, road banks, and so on. Usually the *A horizon*.

Transpiration Water vapor movement through the green surface tissues of plants (see *Stomate*).

Universal soil loss equation (USLE) An equation for predicting, A, the average annual soil loss in mass per unit area per year and is defined as $A = RKLSCP$ where R is the rainfall factor, K is the soil erodibility factor, L is the length of slope, S is the percentage of slope, C is the cropping and management factor, and P is the conservation practice factor.

Vadose zone Aerated region of soil and regolith above a permanent water table.

Value (color) Degree of lightness or darkness of a *color* in relation to a neutral gray scale.

Vascular plants Plants with specialized conductive tissues in organs distinguished as roots, stems, and leaves.

Wasteland Land not suitable for or capable of producing materials or services of value.

Water table The uppermost surface of the *groundwater* or that level in the ground where the water is at atmospheric pressure (i.e., all soil voids are filled with water).

Water table, perched A saturated layer of soil that is separated from an underlying saturated layer by an unsaturated layer.

Weathering All physical and chemical changes produced in rocks at or near the earth's surface by atmospheric agents.

Wetland Transitional area between aquatic and terrestrial ecosystems that is inundated or saturated for long enough periods to produce *hydric soil* and support *hydrophytic vegetation*.

Index

The index contains locators for figures (*referred as* fig.), tables (*referred as* table), notes (*referred as* n.) and boxes (*referred as* box)